El Hacen Ahmed Salem Moctar

COMMUNES OF NOUAKCHOTT - MAURITANIA: BETWEEN DESERT AND TIDAL POOLS

El Hacen Ahmed Salem Moctar

COMMUNES OF NOUAKCHOTT - MAURITANIA: BETWEEN DESERT AND TIDAL POOLS

ScienciaScripts

Imprint

Any brand names and product names mentioned in this book are subject to trademark, brand or patent protection and are trademarks or registered trademarks of their respective holders. The use of brand names, product names, common names, trade names, product descriptions etc. even without a particular marking in this work is in no way to be construed to mean that such names may be regarded as unrestricted in respect of trademark and brand protection legislation and could thus be used by anyone.

Cover image: www.ingimage.com

This book is a translation from the original published under ISBN 978-620-6-71901-4.

Publisher:
Sciencia Scripts
is a trademark of
Dodo Books Indian Ocean Ltd. and OmniScriptum S.R.L publishing group

120 High Road, East Finchley, London, N2 9ED, United Kingdom
Str. Armeneasca 28/1, office 1, Chisinau MD-2012, Republic of Moldova, Europe
Printed at: see last page
ISBN: 978-620-7-96082-8

THE COMMUNES OF NOUAKCHOTT-MAURITANIA: BETWEEN DESERT AND COASTAL PONDS
A: POND IN HALOMORPHIC SOILS WITH SALT CRUST AT THE EDGE
B: INTER-DUNARY POND

C : DEPRESSED POND AND SEBKHAD : BIODIVERSITY POND

JULY 2024, BY, PR DES UNIVERSITES MOCTAR EL HACEN AHMED SALEM

TABLE OF CONTENTS

INTRODUCTION

Nouakchott, the capital of Mauritania, was created ex nihilo in an environment consisting of Sebkhas, shallows and continental dunes, along a very straight coastline. Climatic conditions linked to sea movements and continental seasons have given rise to new landscapes characterised by the outcrop of pools and accumulations of unhealthy water. Today, these landscapes are becoming characteristic of Nouakchott, unlike in the 1970s, when there were just a few sebkhas and marshes along the coast. So we have to ask ourselves why we have reached the stage where floodwaters, ponds and marshes have become a permanent threat to the city, in fact a sort of sword of Damocles hanging over the capital.

THE PURPOSE OF THIS BOOK

This book is based on research carried out as part of the partnership programme between LEERG (the laboratory for environmental studies and geographical research at the University of Nouakchott) and the WACA (West African Coastal Area) project, with the following expected results:

- Inventory outcropping ponds in Nouakchott in 2023 with profile,

- Establish a database on these ponds and their evolution by commune in Nouakchott,
- organise a feedback workshop for the mayors of Nouakchott on t h e basis of an document on ponds.

This book was written by myself, following the workshop held on 15/5/2024 for the mayors of Nouakchott on the profiling of ponds in the capital, which made some important recommendations that will undoubtedly help to better manage and monitor the problem of coastal ponds in a desert capital like Nouakchott.

1 WHEN DID THE PONDS AND SEBKHAS OF TODAY APPEAR ALL OVER NOUAKCHOTT?

1.1 Nouakchott: a breeding ground for marine and rainwater flooding

According to GIZ technical studies (2016/2019), of the 210 km2 of Nouakchott's urban perimeter, 1/3 is located in a flood zone. The topographical level of this flood zone varies from -1 to 0.8 m, i.e. below sea level, or barely above it. The coastline extension mentioned by these same studies is 3 to 4 m in the area of the fish market on the beach, and 20 m in the area of the spur, to the south of the autonomous port of Nouakchott. Lastly, these studies warn that at any time, marine submersions could feed the marshes that frequently occur to the south of this port, posing a danger, particularly to the municipality of El Mina and, beyond that, to the whole of Nouakchott.

1.2 Brief chronology of the main floods in Nouakchott

1916: River flooding cited by references based on local tradition (to be verified), and cited by colonial archives,

1951: Submergence of the embryonic town (Ksar) by the exceptional flooding of the Senegal River,

1983: Submergence of the Aftout Essahéli from Chott Boll to 63 km from Nouakchott,

1987: and on 25 February, a violent storm ruptured the dune cordon to the south of the port of Nouakchott, as soon as the construction work had been completed by the Chinese,

1992: and in August, the sea overflowed following a violent storm and high tides. As a result, the protective cordon was breached by the tides around the Ahmedi Hotel (which was the site from which dune sand was taken for urban construction in Nouakchott),

1997: During the night of 14 to 15 December 1997, a tidal wave broke the barrier beach around the fish market, killing one person. For 10 days, seawater submerged the Aftout depression, between the Sabah hotel and the south of the fish market (Sebkha and Tevragh Zeina municipalities).

2013: Flooding throughout the city, with rainwater raising the level of all the sebkhas and ponds in Nouakchott.

1.3 Continuous outcropping of water since 2001 in Nouakchott, following urban reworking of the soil layers and rising sebkhas and marshes (climate change?)

When did the ponds and sebkhas-mares of Nouakchott first become permanently flooded, as they are today? Was it climate change and the global rise in sea levels? Is it urban pressure and the irrational exploitation of sand and shellfish (which is rarely mentioned)? Is it waterlogging? Is it unsuitable maritime infrastructure such as PANPA (Nouakchott Autonomous Port, known as the Port of Friendship (between China and Mauritania)? All these questions can only be answered on the basis of technical studies and observatories on the Nouakchott coast, accompanied by bathymetric measurements and permanent eco-climatic monitoring. In any case, while palaeontologists (Hebrard, Vernet) have attested to the perennial presence of lagoons to the north of Nouakchott (sebkha Ndrahamcha) and sebkhas in the depressions of L'Aftout, water outcrops were not as present as they are today. The aerial and satellite photos of Nouakchott below, summarised in landscape segments, bear witness to this:

1965: the image shows Nouakchott, with its dunes and vegetation, its shallows with no stagnant water.

1983: The town is invaded by sand in all its communes, including Sebkha and El Mina

2006: No widespread outcropping of water, despite regular rainfall from 2001 to 2006.

2002: Outcropping of ponds at Tevragh Zeina, following the decision
a ban on sand from the barrier beach in favour of continental sand mining.

Photo 1: Landscape segments of Nouakchott in 1965, 1983 and 2006 images:

2 THE CONCEPT OF PONDS, MARSHES, SEBKHA AND SALT PANS

The definition of "mare" can lead to confusion, unless geographical terminology is specifically applied, particularly that relating to coastal forms and landscapes. In Mauritania, there is as yet no geographical lexicon typical of the country's eco-climatic zones, let alone one relating to the coastal zone. However, as part of our preliminary analysis of the profiling of ponds in Nouakchott, we can already distinguish the following forms of outcrop observed: Pond, marsh, sebkha and saline, or rain accumulation zone. In the everyday language of Nouakchott's administrations and population, several expressions are used to describe the outcrop of water in the capital: BURAK, MUSTENGHAAT, MIYAH RAKIDA, SARV SIHI, SEBKHA.... However, it would be necessary to at least distinguish between these types of water outcrops in Nouakchott, according to the geographical terminology used for coastal forms and landscapes:

- **A pond** is a low area covered in water, generally not very large, and fed either by hydrostatic movements of the sea or by other water, such as rainwater or wastewater. The diversity of ponds is therefore linked to the origin of their water, and they can be altered depending on whether they are permanent or temporary. In some countries, artificial ponds are even created for domestic use (salt, laundry, etc.) or for other purposes such as enhancing nature and landscapes. In the case of Nouakchott, the majority of ponds are related to the rising salt water, which prevents any vegetation, except in the case of the emergence of mounds of sand where halophilic plants can grow. The average altitude of the ponds in Nouakchott varies from 0 to 0.8 m. Their hydromorphic soils are formed by the accumulation of sediments (sand, clay, silt and shells). With the presence of salt water and the effect of evaporation, some ponds in Nouakchott have metamorphosed into salt crust concentration ponds (see typology of ponds in Nouakchott). In the long term, and with the continuous outcrop of water in the ponds, we will have muddy soils favourable to some phytoplankton, even attracting migratory and seasonal birds, as in the case of pond F North (see cover page). In this pond, and thanks to the alternation of rainwater and water lost from the SNDE network, typha (a typical freshwater plant) has developed, and has invaded other relic ponds at SOCOGIM PS, and other areas in Nouakchott.

- **Marshes** are almost a continuation of coastal landscapes, and in geography we speak of coastal marshes and maritime marshes. However, marshes can be freshwater if they are located at the mouth of a river flowing into the sea. Marshes, such as the one south of the port of Nouakchott, are subject to the rhythm of high tides and spring tides. The marshes are constantly being created on the coastal edge of the seas, and are therefore a function of tidal movements. The tides are subject to the attraction of the moon and sun on the sea, but also to the rotation of the earth. Tides can be low or high, and can turn into tidal waves like the one in December 1997, which killed one person at the fish market in Nouakchott. If the tidal wave is exceptional, it can also turn into a disaster, like the one observed in 1983, which submerged the entire coastal strip of Aftout Essahéli for two years, threatening Nouakchott 63 km along the Rosso/Nouakchott road. Naturally, such a tidal wave could submerge at any moment and invade the capital's neighbourhoods, thus raising the level of the ponds already present in practically all of Nouakchott's communes (even Toujounine, the most continental of the communes).

- **Sebkhas are** defined as closed depressions when they are located in continental areas far from the sea. On the other hand, coastal sebkhas close to the sea are said to be open because they are in contact with the sea, either continuously or discontinuously (when, for example, underground contact with the sea is interrupted by mounds of sand such as the barrier beach). The sebkhas of Nouakchott are therefore coastal sebkhas, as are the sebkhas that surround them **from close by**, such as the sebkha of Ndrahamcha (whose west/east contours start from Jreida and the new Nouakchott airport), **or from afar,** such as the south-Aftout sebkhas of Jdeir (25 km south of Nouakchott) and Tefourtess (62 km further south). The belt of sebkhas continues even further south, along the Essahéli Aftout, with sebkhas that are larger in terms of surface area and quantity of salts, such as Oumlekhcheb, Agamoun, Tinijmara and Moiijeran. It should be pointed out that certain coastal sebkhas corresponding to breaches (mentioned on 6 January 1951/in an IFAN mission during the flooding of Nouakchott by the Senegal River) in the **vicinity and in Nouakchott town may be sebkhas with shifting soils.**

Some of the ponds in Nouakchott are in fact the remains of sebkhas that have been poorly identified by urban planners and the residents who build there, (See LEERG puddle inventory survey sheet in the appendices).

The sebkhas of Nouakchott-ville and the surrounding area are outcropped either without salt or with salt:

. Those that are salt-free, for example, are often located between the dunes, and the sand therefore limits the concentration of salt in this type of sebkha. This is the case, for example, of the Soukouk sebkha located between F Nord and Ain Talh.

. As for the sebkhas in the depression of L'Aftout Essahéli, in the communes of El Mina, Sebkha and Tevragh Zeina-west, they are all located in halomorphic clay soil and, with evaporation, they generate a considerable saturation of salt crystals, much to the delight of a few salt producers, who sell the recovered salt throughout the town and inland. This type of sebkhas is easy to identify, as the blistering salt covers the entire surface and then solidifies the depression into a loose clay-saline layer. This does not prevent some people in Nouakchott from building on sebkha soils, and after noticing the effects of the salinity on their buildings, they completely abandon their expensive houses (see photos in this report).

- Salt flats are sebkhas identified with the presence of fossil salts (salt bars such as at the Idjil sebkha near Zouerate, or the Nterert sebkha, between continental dunes near Tiguent) or salts concentrated by the renewal of water from the sea or rain. Salt bars could not therefore have formed in Nouakchott, as it turned out that all the sebkhas are apparently in contact with seawater (which remains to be verified through a much-needed study of the sebkhas along the Mauritanian coast). As for the Ndrahamcha sebkha north of Nouakchott, the concentration of sediments in the sebkha (in Ass. Sénégal. Et. Quatern. Afr., Bull. Liaison, Sénégal, no 4'9, décembre 1976 géochimiques préliminaires sur la sebkha de Ndrhamcha / Mauritanie. par MAGLIOHE et CARH) is composed of minerals, such as gypsum added to salts, of lesser importance. (See also the thesis by Professor Sabar of the University of Nouakchott on the Ndrahmacha sebkha). However, it has been proven that seawater rises directly towards this sebkha, particularly around the village of Tiwilitt (see Figure 1):

Figure 1: Cross-section of the Ndrahamcha sebkha (source: MAGLIOHE and CARH/1976)

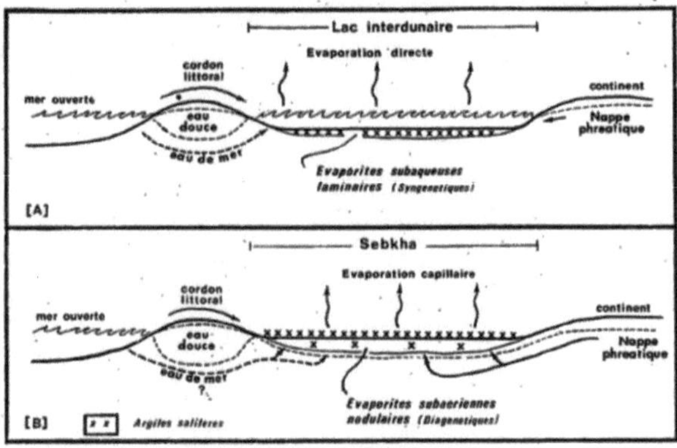

3 NOUAKCHOTT IN A LAND OF SEBKHAS AND DEPRESSIONS, HENCE SOIL SATURATION

3.1 Geology and geomorphology of the communes of Nouakchott

The communes of Nouakchott are located in the geological basin known as the Senegalo-Mauritanian basin on the West African Atlantic coast. This basin is made up of sedimentary rocks characterised by the presence of sand, clay and limestone (see map and geological section below). In the eastern part of this basin, the Trarza aquifer has formed, the geological boundary of the saline wedge around 50 km from the sea. The saline depressions of Aftout Essahéli and its sebkhas were formed during the Quaternary, and were then separated from the sea by the present coastal dune belt. The aridity and irregular groundwater supply of these depressions has led to evaporation, which has increased salinity. The salt content is such that halophilic vegetation has difficulty in proliferating, except in the event of heavy rainfall.

Map 1: Geological section of Nouakchott (Source: Bassin-Sénégalo-Mauritanien/AIEA/2017)

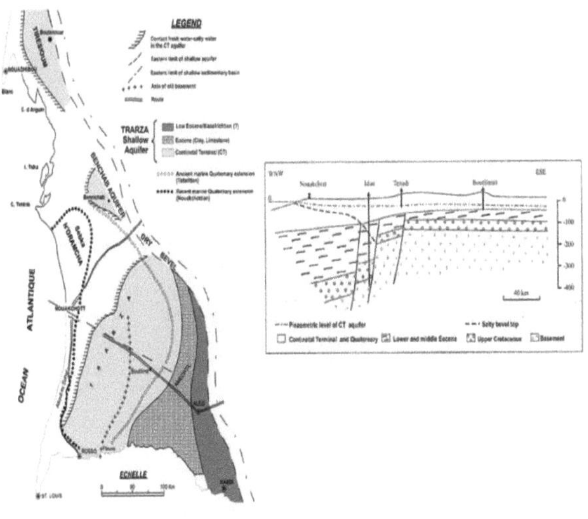

11

EVOLUTION OF THE MORPHO-URBAN SPACE 1908/2021 RECONSTITUTED FOR THE SEAFRONT MUNICIPALITIES :

The two morphological sections of Nouakchott, in 1908 (Gruvel-Chudeau) and 2021 (LEERG), taken towards the ocean, show how the dune complexes cited and photographed by the Gruvel mission have disappeared under urban pressure. Their location has changed, giving way to sebkhas or outcropping ponds here and there, in the middle of the urban districts of El Mina and Sebkha. The ponds and ponds-sebkhas that are currently outcropping have become Nouakchott's new defining landscapes. Coastal dunes are disappearing due to coastal erosion and human activity.

Figure 2 series:Schematic compared of Nouakchott 1908/2021(source: Author based on Mission Gruvel 1908) :

1908 :

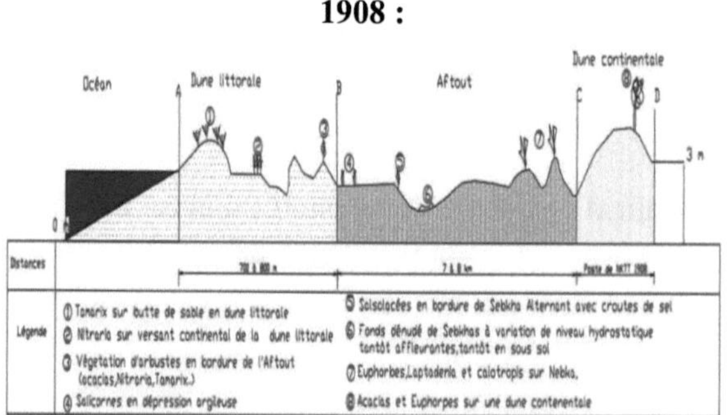

Coupe schématique de Nouakchott 1908 (A. Gruvel) mise en forme par CMF 2022

1921:

Coupe schématique de Nouakchott 2021, réalisée par CMF 2021

3.2 Type of soil at low altitude favouring the outcrop of ponds in Nouakchott

The shallow soil types in Nouakchott favour the outcrop of groundwater and the sebkhas, which are in contact with the sea (see cross-sections). 57% of Nouakchott's perimeter is covered by this type of soil, which exposes the capital to the hazards of submersion and pluvial flooding. The altitudes of these soils vary (see map 2) in the seafront area (Sebkha, El Mina and Tevragh Zeina municipalities) from -3m to 1.5m. As for the municipalities located on the continental front (Toujounine, Arafat, Riadh-Est), the altitude of the soil varies from 1.7 m to 11 m, towards Ouad Naga in the east.

Map 2: Soil map and altitude in Nouakchott (source: Jica/Sdau 2018)

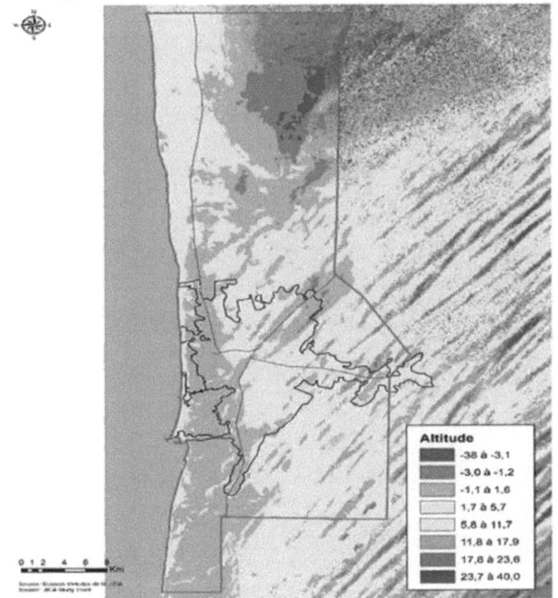

3.3 Soil in the communes of Nouakchott threatened by submersion and flooding

Ponds and sebkhas of varying depths are found in the lowest areas (see characteristics of ponds by commune). When rainfall exceeds 40 mm, for example, the ponds overflow, encouraging the spread of water-borne diseases. Flooding is a real ordeal for the people of Nouakchott. Depending on the topography of each commune, some ponds remain stagnant throughout the year, causing continual inconvenience to local people, who do not hesitate to flood them with rubbish. Map 3 below and the accompanying schematic section show the contour lines and a general distribution of ponds in each commune of Nouakchott. The GIZ 2016/2019 studies have shown that the sea, which is in contact with these ponds and sebkhas, will rise in the coming years by an average of 0.2 to 1 m. This will pose a serious threat to the communes of Nouakchott, especially in the context of the climate change forecast. The same studies have also shown that the coastline is advancing from the sea towards the mainland, i.e. by 3m in ten years (from 2004 to 2014).

Map 3: Distribution and schematic cross-section of ponds in the communes of Nouakchott (source: Jica/Sdau 2018)

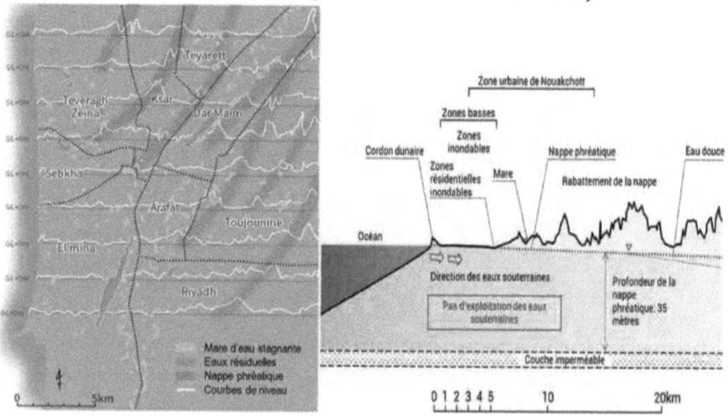

4 METHODOLOGY FOR INVENTORYING PONDS IN NOUAKCHOTT 2023

4.1 Selection criteria for ponds inventoried by LEERG teams

The level at which the marshes, pools and sebkhas fill up depends on the hydrostatic level and the amount of rainfall recorded in Nouakchott, but also on the effect of underlying marine incursions. Management of these outcropping waters, whatever their origin, is becoming a problem for the administrative and municipal authorities, who have no other solution than to fill in these vast expanses of sand and clay, thus taking up land that is already fragile and vulnerable on the outskirts of Nouakchott. Our inventory methodology is based on the following main references, added to the bibliographical references at the end of this book:

- The IFAN report produced on 6 January **1951** during an aerial mission on the flooding of Nouakchott, which refers **to sebkhas with shifting soils around Nouakchott.**
- Report on morphological dynamics in Nouakchott (Barrére **1983**),
- GIZ studies (ACCCV Project) in **2016/2019,**
- JICA studies on the Nouakchott SDAU in **2018,**
- The ACF/LEERG surveys carried out in **2019** on flooding in three communes of Nouakchott (Sebkha, El Mina, and Dar Naim), with the experimentation of avaloirs as solutions to wastewater, added to surveys on the perception of flooding by households in the communes surveyed,
- The AREDDUN(Region) project, which carried out the study: "the Nouakchott region in the face of climate change **2019".**
- Report on the first survey of ponds in Nouakchott carried out by LEERG teams on **28/12/2020.**
- The doctoral thesis on "Coastal governance and climate change in Mauritania" defended by Mr Fadel in December **2023** at the University of Lille in France.

The ponds that are the subject of this book are selected from all the communes of Nouakchott, in accordance with the WACA/LEERG research programme, which began in November 2023. The methodological process consists of :

- Geographically locate each waterhole using GPS, within the urban perimeter of Nouakchott and not the peri-urban perimeter,

- View each waterhole using local photos or Google photos,

- Characterise each pond, its uses, habitats and immediate environment,
- Present the data available while awaiting the arrival of the probe and the drone promised by the WACA project.

- Finally, the presence of birds in certain ponds should be noted, especially during the months of December 2023 and January 2024, when birds migrate from Europe.

4.2 Period to report

It should be noted that the inventory period for the ponds began on 11 November 2023, at the end of the wintering period, and with a view to the winter solstice (21/12/2023), a period favourable to marine currents, storms and high tides on the Mauritanian coast. It ended in February 2024.

4.3 The profiling and characterisation sheet for the ponds surveyed in Nouakchott (see appendix)
The "Pool profiling" sheet has several sections:

-   Information on the site of the pond and its muddy, sandy and clayey   soils.
-   The origin of the pond: marine, pluvial, man-made, created following network leaks (SNDE, ONAS) or removal of the surface layer of sand or shells for urban   construction purposes.
- The pond and its surrounding environment: fauna, amphibians, etc.
- The landscapes of the pond: vegetated, as a body of water, or infested with rubbish,
- Domestic uses and land ownership,
- Local information on the pond and additional data from residents,
- Threats and inconveniences caused by the presence of the pond, or on the contrary, useful, like landscape in water level, to look at for some, as in Tarhil (Commune Riadh)

At this stage of the fieldwork, we have not prioritised the ponds in Nouakchott, particularly in terms of hydrological, heritage and eco-tourism criteria. This will certainly come up in our next studies.

4.4 Database of ponds in Nouakchott

On the basis of the profiling form, an initial database of ponds has been compiled, with a GPS location that links (in Excel) to a form filled in by the teams for each communal pond (see appendices).

4.5 Mapping adopted for the inventory of ponds

The ponds identified in this first phase are configured on a map by commune, and on a Google urban photo base. Two types of geo-referencing are applied by the teams in the field: GPS pointing and map pointing.

4.6 Applied parameters

At this stage, pending the arrival of the probe and the drone, we will only present standard data on conductivity, temperature and pH in the ponds surveyed. As a reminder, for conductivity, the following comparative references should be considered:

- Conductivity and temperature measurements are always taken at the same time, because temperature influences the conductivity measured, day and night. Temperature has a major influence on water conductivity. To compare conductivity values from one season to another and from one body of water to another, they need to be calibrated to a water temperature of over 20°C. Once adjusted, they become specific conductivity data.
- The conductivity of tap water in Nouakchott (SNDE), for example, averages 0.2 μs/cm (micro siemens per metre).
- The conductivity of pure water averages 0.055 at a temperature of 25 degrees.
- The conductivity of seawater is around 50 m/s at 20°C.

The conductivity of brackish water is around 1,000 to 10,000 μS/cm at 20°C.

The brackish water in the ponds in Nouakchott is derived mainly from the rising saline groundwater near the sea. This water becomes conductive due to dissolved substances such as sodium chloride. However, the conductivity of the water in some ponds is attenuated by other sources of supply, such as the SNDE or ONAS wastewater networks. During the winter season, and depending on Depending on the amount of rainfall, conductivity can also drop in value, sometimes reaching 800.0 μS/cm) at 27°C. We will present average conductivities that are characteristic of ponds in Nouakchott, but which vary greatly by commune when they are in seafront or continental fringe areas. The lowest average conductivity values in the ponds in Nouakchott are found during the cold season, i.e. on average 100 to 250 µs/cm, with temperatures varying between 18 and 28 degrees. The highest average conductivity values in the ponds of Nouakchott in summer are 190 to 700 µs/cm, with temperatures varying between 27 and 34 degrees.

PH: this is defined as the hydrogen potential in water. It is also an important measure of water alkalinity. It can be neutral when it is equal to 7. Below that, it is acidic, and therefore not conducive to the development of certain plants, and above that, it is considered basic. The average PH in the ponds in Nouakchott was measured at 7.5 to 9 in the cold season, and 8 to 9.5 in the hot season.

4.7 Number of ponds per municipality according to preliminary inventory November 2023

An exhaustive inventory of ponds in Nouakchott requires an ongoing programme of counting, particularly by season (especially during the spring and rainy seasons). Nevertheless, this initial inventory will give an idea of the number of ponds in the communes of Nouakchott.

NUMBER OF MARES PER COMMUNE AND IN ORDER OF INVENTORY (November 2023 to January 2024):

DAR NAIM :105

TV ZEINA 26

SEBKHA22
EL MINA 14

RIADH 9

ARAFAT 2

TEYARETT : 6
KSAR : 4

TOUJOUNINE: 0

5 TYPOLOGY OF SEAFRONT PONDS IN NOUAKCHOTT

In Nouakchott, a distinction must be made between :

- A low-lying area flooded by rain or a water leak in the network, with the ground saturated at the slightest drop.

- A pond rising from a water table close to the average level of the topographic surface at Nouakchott, with 0 to 0.8 m (at around 0.5 to 0.8 m).

- A sebkha that rises with the moon's tides or a good night's s l e e p .
rainfall,

- An artificial pond created by digging up the ground to meet the needs of roads or urban construction.

5.1 TYPES OF PONDS IN THE MUNICIPALITIES ALONG THE NOUAKCHOTT COASTLINE: EL MINA, SEBKHA, AND TEVRAGH ZEINA

Most of the ponds in these three communes, known as "coastal communes", are located in the Aftout Essahéli depression adjacent to the coastal strip. These communes alone account for 40% of Nouakchott's population, but they also have the distinction of containing several infrastructures of national importance to the country, such as the port and its fuel depots, domestic gas storage tanks, the country's flour mills and cement works, and many other infrastructures. The municipality of El Mina is also known for its seasonal marshes, a phenomenon it shares with the municipality of Sebkha, particularly during periods of high water and high tides.

Communes along the Nouakchott coastline :

Name of the municipality	characterized as	Surface area in Km2	Population 2021
TEVRAGHZEINA (NKTT)	Urban	30	67382
SEBKHA(NKTT)	Urban	14	64 316
El MINA(NKTT)	Urban	90	154 963

5.1.1 Typology of ponds at El Mina

The El Mina marshes, particularly those to the south of the PANPA, can sometimes turn into swampy marshes, threatening the nearby Dar El Beidha area and the industrial concessions around the perimeter of the port. These marshes, which vary in size, are regularly submerged by marine submersion, encouraged by the disappearance of the barrier beach in this part of the coastline, as well as by the inappropriate construction of the port of Nouakchott and its jetty. In winter, rainwater submerges these ponds, and in summer, they dry out and become covered in saline efflorescences, as is also the case in other ponds in the commune, which are used for salt production by local people (Photo 2 below):

Photo 2: Type of pond at El Mina, known as the "salt pond

This type of marshland is being nibbled away at by the housing estates allocated by the State, and the beneficiaries of these plots in poor neighbourhoods are doing everything they can to fill in the salty land and build on it. It's obvious then that their habitat is in a risk zone, and could therefore at any time be subject to marine fluctuations (the sea is 5 km away) and occasional heavy rains.

Photo 3: Type of pond known as: "Pond resulting from the clandestine (sometimes nocturnal) digging of a shell quarry" (They can also be distinguished as a type of artificial pond).

These types of quarry are dangerous, as they encourage flooding and landslides. The risks are even greater when these quarries are close to the coastline, as shown here in photo 3 (the boats in the port of Nouakchott can be seen in the background).

In the eastern part of the municipality of El Mina, all the low-lying areas at 0 metre level in the Aftout Essahéli maritime depression have been submerged by urban development. In the cross-section of Nouakchott drawn up in 1908 by Gruvel/Chudeau, these areas, which are now heavily urbanised, are shown in the diagram as areas of pluvial flooding, sometimes alternating with outcropping sebkhas. As a result, rainwater and rising water tables are now to be found in the courtyards of the houses in El Mina, whose walls and foundations are crumbling and surrounded by thin films of salty damp. In these parts of El Mina, "Individual sanitation is also difficult, because digging pits comes up against the proximity of salt water" (SNEIH 2006), which is one metre deep.

Photos 4 in series: Type of pools submerged by rubbish :

Ponds at Bagdad and Socogim Ps: Typha and pond water bodies submerged in turn by rubbish (1and 3)/Other Type of pond: (Made in): Used oil dumping ground (Arab High School and concessions for garage and Ministry of Hydraulics shops + ONSER)

5.1.1.1 Preliminary inventory of El Mina ponds November 2023

If the ponds inventoried at this stage in El Mina are smaller than those in Sebkha and Tevragh Zeina, this is due to the urban explosion, which has submerged the entire L'Aftout Essahéli depression, and as a result, most of the ponds that were outcrops at the end of the 90s in El Mina are now under urban construction. As for the Socogim PS dwellings, they were built squarely on a sebkha, which until the 1980s was known as Sebkhit El Kebaa. On this inventory map, the ponds outcropping in November 2023 are highlighted in red, on the western outskirts of the municipality, close to the coastline, and still unoccupied by urban development. This is due to the persistent existence of ponds in this area, added to the bitter experience of the MELAH allotment awarded in 1993 by the Wilaya of Nouakchott. The Wilaya deliberately overlooked the risk of flooding, which transformed the area allocated by the extension of El Mina into veritable salt ponds during the winter of 1993. The ordeal at the time was such that the State cancelled the allotments and transferred MELAH to the eastern dunes of Nouakchott (to this day, the district is called MELAH, even after it was transferred to the dunes in the Toujounine commune).

Map 4: Preliminary inventory of ponds at El Mina in November 2023 (source: Leerg on Waca 2023/2024 research programme)

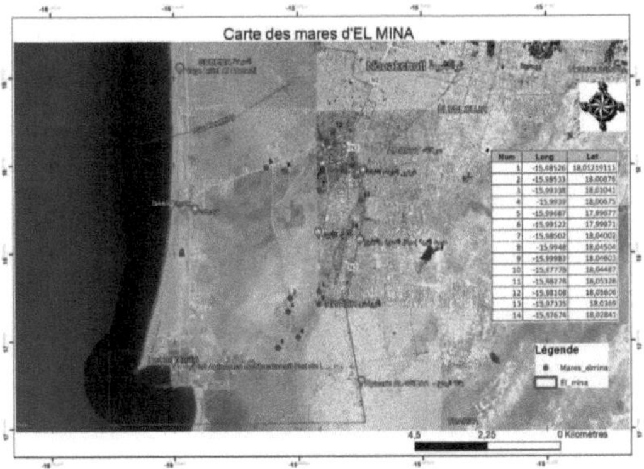

5.1.1.2 Profile of ponds at El Mina

The El Mina pools run parallel to the dune belt and are located in the middle of the Aftout Essahéli depression, with water levels of 0 to 0.5 m, and even - 1 m in some places. Their outcrops depend on sea variations (storms, high tides and surges, submersion, etc.). The GIZ 2019 studies have shown that beyond the barrier beach, there are underground exchanges between the sea and neighbouring land, such as the Aftoutienne depression, where the communes of Sebkha and El Mina are located, and even part of the commune of Tevragh Zeina. The gradual rise in sea level favours the intrusion of water, and its seasonal outcrop in the ponds of the commune. The peaks in the outcrop of water in the ponds in the seafront communes are observed from January to April (surges and spring tides), and from July to October, depending on the amount of rainwater. The ponds and marshes at El Mina have formed in flat soils with little vegetation. The shallows of these pools are characterised by areas of sediment accumulation (sand, clay, silt and shells). Over time, and with the continuous outcrop of brackish water in these areas, we will have muddy or marshy soils.

Photo 5: El Mina used to have dune ridges in 1983, but these have since been erased by building work, which has caused pools to emerge as a result of hydrostatic pressure (Crêtes vives du 6e arrondissement/ Cliché P. Barrère) In the background, the Moroccan mosque and its market.

5.1.1.3 Landscape features of the El Mina ponds :

Maximum water depth measured on average in 3 ponds: O to 1m 20, mainly from past or present seawater infiltration during submersion, storms, surges or tidal surges (rainwater is seasonal and depends on quantities).

Deep soil: Accumulation of sediments (sand, clay, silt and shells) + salts absorbed on clay at the edge of the pond.

Soil around ponds: on the coastal fringe, the soil is becoming slippery, particularly around the Hotel Ahmedi and around the French Navy camp. These soils become even more fragile in the event of a good winter.

Highest average conductivity, recorded in summer, in the ponds of Marbatt and Dar Beidha-Samine: 190 to 700 µs/cm, with temperatures varying from 27 to 34 degrees.

Slightly sloping banks: O to 25% on average,

Anthropogenic waste: rubble, plastic, animal carcasses, used oil, septic tank waste.

Disfigured landscapes: as a result of shellfish and alluvial quarrying, sand has almost disappeared from the commune's landscapes, except in the coastal dunes, which are subject to clandestine sand extraction transported in carts, or in bags hidden in private cars.

Pisces: probably not

Invasive species: rare, except for typha at Socogim Ps, as the water is mixed with domestic wastewater and accumulated rainfall,

Main use of ponds: salt collection

Drainage: exceptional pumping by ONAS (notably in Dar El Beidha, September 2023), civil security and military engineering, in other urbanised puddle sites, and at crossroads of the commune's alleyways.

Ponds in El Mina causing climate refugees: In 2014, the LEERG survey recorded 38 households who had abandoned Socogim Ps and moved to Riad and Arafat: either through the purchase of new land or through renting. In 2020, the ACF/FISONG survey recorded 72 cases of house abandonment in El Mina, due to the pressure of floodwater from ponds and sebkhas, or following the 2013/2014 rainfall.

5.1.2 Typology of ponds at Sebkha

As its name suggests, the commune of Sebkha was built on a field of sebkhas, which some geologists even link to the great sebkha of Ndrahamcha (or dried-up lagoon), 50 km north of Nouakchott. It is also true that when the fifth arrondissement (or fifth) was built, the ancestor of the Sebkha Moughataa, the risks posed by the sebkhas were forgotten. The tradition of the Bouhoubeiny coastal nomads teaches us that you should never build a dwelling on a sebkha, even if it is a Khaima. The traditional maxim also teaches us that you should never approach a sebkha when it is boiling, or when its water is being carried away by very strong winds. Today, the city dwellers of Nouakchott live in the commune of Sebkha and have to contend with the inconveniences caused by the capillary rise of salt, which is very present in the salt depressions. The particularity of the commune lies in the fact that the outcrops of ponds are more related to the rising sebkha than to pluvial flooding. Of course, when heavy rainfall is added to the sebkha water, it becomes unbearable for the inhabitants of the Moughataa, who cry out that their commune is a "water hostage" (CRIDEM media website, 23/8/2022). In the street, the citizens of Sebkha are also hampered, especially when the evaporation of the water is becoming intense, and as a result, the salt is rapidly crystallising, preventing cars from circulating. All areas of the commune are now unfit for construction, as the salt attacks the walls and the brackish water seeps into the individual household sewage systems. Some people are abandoning their homes, or selling them to foreigners, who then become temporary residents of Sebkha, just long enough for these migrants to find work, or just long enough for their projects to succeed. In 2020, the ACF/LEERG survey revealed that more than 1% of houses had been abandoned in the commune of Sebkha. In any case, we look forward to the application of the DAL (Coastal Development Directives, contained in the Master Plan for Coastal Development) in the commune of Sebkha, which "firmly prohibits construction in highly flood-prone areas of the sebkhas".

TYPE of pond: Sabah Hospital pond or sebkha?

The Sebkha salt pans (photo 6) are regularly scoured to such an extent that one wonders whether this is the great salt works of Nouakchott, which the first explorers reported in the 18th century. In this sebkha-mare, the water level is 0.2 m, and salt accumulates constantly during the hot seasons, resulting in a mud that is sometimes pink, sometimes white with salt (see photo 6). When questioned, the salt workers we met said that the sebkha at the Sabah hospital is fed regularly and underground by the sea, which is 7 km away. The surrounding sebkhas-mares only produce salt after the winter season.

Photo 6: Salt from the Mare-sebkha at Sabah Hospital

Photos 7 in series: comparative photos of pond types: ATOIT pond or the hydro system of an urbanised sebkha.

The nature of the soil, proximity to the sea and rapid urbanisation have transformed the sebkha-mare known as Atoit into a veritable sub-saline district. The city dwellers who chose to settle in Atoit probably preferred to live close to the sea. the sea and the fish market. LEERG's monitoring of this sebkha pond since 2002 shows that urban occupation accelerated from 2010 onwards (see photos comparing 2002 to 2011 below).

ATOIT 9/8/2002: salt recovery area and rubbish dump

ATOIT 21/5/2004: the salt depression is drained and SOCOGIM begins a backfilling programme to install fences made of rehabilitated walls.

ATOIT 5/6/2010: Urban pressure and transformation of the SOCOGIM fences by the beneficiaries into villas, medium and high standing/Remediation of certain parts by the military engineers

ATOIT 8/10/2011: Pumping by the Ministry of Hydraulics (ONAS) with no result until the arrival of the Chinese rainwater and stagnant water purification project.

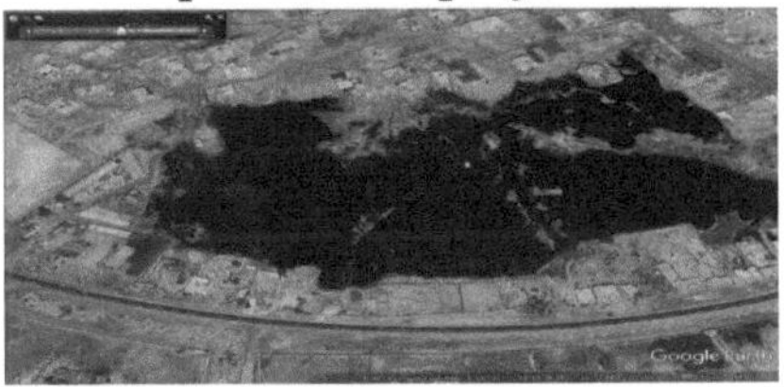

ATOIT 14/12/2019 : Slight reduction in the volume of the pond following the pumping systems installed by ONAS and the Chinese from 2016 onwards

5.1.2.1 Preliminary inventory of Sebkha ponds November 2023

Sebkha ponds are scattered throughout the municipality, and the extent of their outcrop is a function of urban pressure and topography (in particular the lowest points on the ground in alleyways or public spaces). However, sebkha pools tend to outcrop much more close to the barrier beach (see map 5 of sebkha pools). The presumed contact of these sebkhas with the sea is almost irreversible (even though there are no studies on sebkhas and their underground functioning in Nouakchott). In addition, the 2017 IAEA studies showed that the salt wedge is at the IDINI border, 60 km east of the capital. Conductivity in the ponds-sebkhas recorded over 3 ponds averages 1,500 to 10,000 µS/cm at 28°C. During the rainy season, flooding raises the level of the sebkhas, and at the end of the rainy season, the water dries out, and with evaporation and high temperatures, the salt submerges the commune of Sebkha. Several levels of the questionnaire are still missing in the commune of Sebkha, as there are headings such as land ownership status, in particular lots regularly submerged by water, which require surveys. and gathering information from government departments local.

Map 5: Preliminary inventory of ponds at Sebkha in November 2023 (source: Leerg on Waca 2023/2024 research programme)

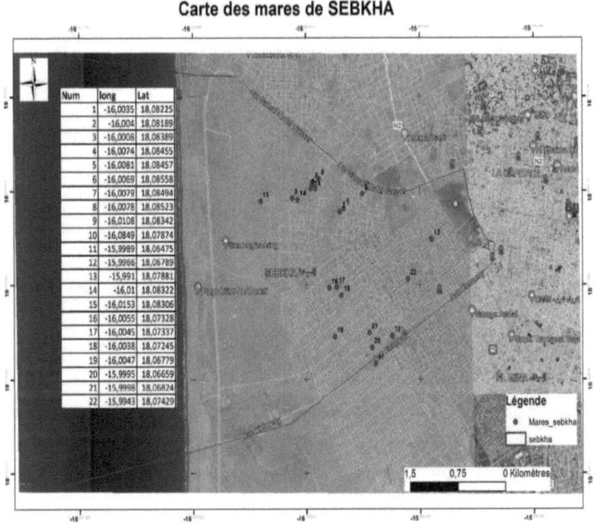

Num	long	Lat
1	-16,0035	18,08225
2	-16,004	18,08189
3	-16,0008	18,08389
4	-16,0074	18,08455
5	-16,0081	18,08457
6	-16,0069	18,08558
7	-16,0079	18,08494
8	-16,0078	18,08523
9	-16,0108	18,08342
10	-16,0849	18,07874
11	-15,9989	18,06475
12	-15,9966	18,06789
13	-15,991	18,07881
14	-16,01	18,08322
15	-16,0153	18,08306
16	-16,0055	18,07328
17	-16,0045	18,07337
18	-16,0038	18,07245
19	-16,0047	18,06779
20	-15,9995	18,06659
21	-15,9998	18,06824
22	-15,9943	18,07429

Salt is so important that households in Sebkha buy it locally for their cooking needs and in front of their homes. Photo 8 below shows the concentration of salt in the plots around the villas that have already been built.

Photo 8: Concentration of salts around vacant lots at Socogim-plage (source: Auteur 2023)

The inventory of sebkha ponds revealed four methods used by local residents to protect the foundations of their endangered villas. Despite this, the result is the gradual abandonment of the houses shown in photo number 10.

Photos 9 in series: photos of the methods used by the people of Nouakchott to combat rising salinity:

Process 1: (photos A and B below): Backfilling the lot with sand (which is permeable to rising sebkhas) (Source: Author 2023)

A:B :

Process 2: Embankment of the foundation with tarmac to temporarily prevent salt build-up.

Process3 : Stonework of the lot in rocks continental from the interior of the country (Aioun, Atar)

Process 4: Tiling the foundation and fence, which does not prevent wear and tear on walls due to the effect of saline capillary rise.

After processes 1, 2, 3 and 4: Photo 10: At the end of the day, the houses in Sebkha were abandoned, despite the protective measures taken. See Climate refugees in Nouakchott

5.1.2.2   Profile of ponds at Sebkha

As with the ponds at El Mina, the main ponds at Sebkha are also located parallel to the barrier beach, in the middle of the Aftout Essahéli depression, with water levels ranging from 0 to -1 m. The ponds in the commune have benefited from Chinese pumping, which has had a temporary impact on households located near perennial sebkhas. This has slightly reduced the periodic cycle of emptying septic tanks, a real ordeal for households in Sebkha. However, the dominant morphology in the commune,remains a type of pond known as the Sebkha pond. If the profile of the ponds at Sebkha requires continuous pumping, this could have unforeseeable consequences, such as soil collapse and landslides on land already occupied by thousands of households. At Socogim- plage, the State has filled in the sebkhas, but this cannot be a solution because of the pressure it will exert on the already vulnerable water tables, which will have repercussions on other areas, aggravating the rise in groundwater levels.

5.1.2.3 Landscape features of the Sebkha ponds

Maximum water depth measured on average in 3 sebkhas ponds: O to **0.8 with stagnant sebkha water.**

Soil at depth: sandy-clay sediment, mud around the edges of sebkhas

Soil around ponds: High concentration of salts

Highest average conductivity, recorded in October 2023 in 3 exclusive sebkhas: 300 µs/cm, with temperatures varying from 29 to 34 degrees. Conductivity in the sebkhas pools recorded in 3 pools averaged 1,500 to 10,000 µS/cm at 28°C in November 2023.

PH: The average pH varies between 8.4 and 9.5, similar to slightly basic seawater.

Slightly sloping banks: O to 0.5% on average,

Presence of man-made waste: rubbish dumps

Landscapes of sebkhas, and in winter, the water surfaces are abundant, hampering the urban economy and car travel.

Aquatic life: a few rare dragonflies and butterflies.

Invasive species: Tamarix in certain houses, calotropis in individuals, Zygophyllum waterlotii

Main use of the sebkhas ponds: salt collection

Wastewater treatment: Chinese pumping from 2016 to 2022

Mares-sebkhas having led to houses being abandoned (figures still to be researched):

SUMMARY OF THE HYDROSYSTEM OF THE MAIN SEBKHAS IN THE COMMUNE OF SEBKHA :

Name of the continuous or discontinuous area of a series of pools on sebkhas bottoms	Geographical coordinates of ponds and sebkhas in the commune of Sebkha		Volume of water saline, with intensity in colour (Importance from blue to beige)
	Latitude	Longitude	
Cité plage (concentrated salt)	18°05' 12,3"	16° 00' 09"	
Concorde area and Ministère pétrole (greyish salt crust)	18°05' 11,3"	15° 59' 51,1"	
Basra and Kouva (salt in a surface crust very mixed with clay)	18°04' 49,9"	16° 00' 15,0"	

5.1.3 Typology of ponds in Tevragh Zeina

There are two types of ponds in the commune of Tevragh Zeina:

- Firstly, there are the pools in the western part of the municipality, which are in the middle of the L'Aftout salt depression, extending towards the large Ndrahamcha sebkha (which begins at the new Nouakchott airport). The characteristics of these pools are similar to the sebkha pools already described in the communes of El Mina and Sebkha.

- And then there are the ponds in the northern and central part of the commune, which are characterised by the fact that they were under a dune belt before the dunes that protected them were reworked or stripped away by sand trucks, as in the case of the Soukouk pond (monitored by LEERG from 2002 to 2023) below:

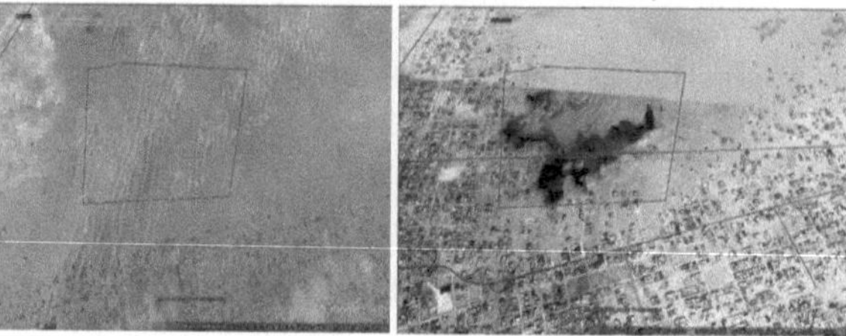

Pond in the SOUKOUK area: covered by the dune cordon in **2002**, then lorries extracted the sand, and **from 2007 to 2019**, the pond has emerged to the present day **(November 2023)** despite the work of private property developers, who continue to pay every lorry to deposit in the pond the rubble from demolished houses somewhere in Nouakchott. By backfilling the pond in this way, they build flats and houses, whose **tenants quickly become disillusioned** as soon as they move in, particularly with the daily filling of septic tanks as a result of groundwater pressure and capillary rise on the walls of the houses (NB: this pond was also the site of all the rubble and breakage from the Senate building, demolished in 2017/2018).

5.1.3.1 Preliminary inventory of ponds in Tevragh Zeina November 2023

On the inventory map, we can clearly see the Nouadhibou road which separates the two types of pond described above. In the eastern part of this road, the continental dunes interfere with the urban districts of the commune, momentarily blocking the extension of the shallows submerged by the pools, as shown in photo 13 :

Photo 13: Pond held back by continental dunes in Nouakchott

The inventory also showed other types of ponds that had benefited from connections and water leaks, such as the lost ONAS or SNDE networks; this has encouraged biodiversity and typha vegetation, to the extent that it looks like the Senegal River in southern Mauritania (see photo 14 below):

Photo 14: Typha vegetation in Nouakchott, favoured by a soil alternating between fresh and salt water/At the heart of this vegetation, biodiversity and refuges for seasonal and migratory birds develop.

The ponds of Tevragh Zeina are constantly being filled in for the construction of luxury villas, and because of the land value of the plots allocated by the State, in the very heart of the pond, or in its vicinity. City dwellers in the commune of Tevragh Zeina also have a certain standard of living, which enables them to afford the services of lorries and graders, unlike the inhabitants of the El Mina and Sebkha areas.

Map 6: Preliminary inventory of ponds in Tevragh Zeina in November 2023 (source: Leerg on Waca 2023/2024 research programme)

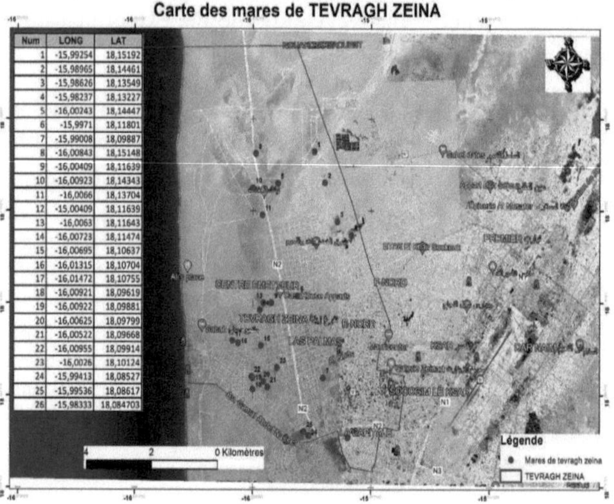

Num	LONG	LAT
1	-15,99254	18,15192
2	-15,98965	18,14461
3	-15,98626	18,13549
4	-15,98237	18,13227
5	-16,00243	18,14447
6	-15,9971	18,11801
7	-15,99008	18,09887
8	-16,00843	18,15148
9	-16,00409	18,11639
10	-16,00923	18,14343
11	-16,0066	18,13704
12	-15,00409	18,11639
13	-16,0063	18,11643
14	-16,00723	18,11474
15	-16,00695	18,10637
16	-16,01315	18,10704
17	-16,01472	18,10755
18	-16,00921	18,09619
19	-16,00922	18,09881
20	-16,00625	18,09799
21	-16,00522	18,09668
22	-16,00955	18,09914
23	-16,0026	18,10124
24	-15,99413	18,08527
25	-15,99536	18,08617
26	-15,98333	18,084703

5.1.3.2 Profile of ponds in Tevragh Zeina

The pools in the eastern part of Tevragh Zeina are surrounded by sand dunes up to 3m high, but with a very gentle slope (1.5%). The presence of sand, combined with frequent dews and fogs, has encouraged semi-desert coastal vegetation, particularly in the ponds of Soukouk and F Nord: Leptedania, calotropis, Waterlotti,..... The presence of sand dunes has not, however, alleviated the frequent outcropping of brackish water in these pools, and there is still the problem of the sustainability of this water, particularly its underground supply. The danger in Tevragh Zeina is that most of the public infrastructure (schools, ministries, university, etc.) has been built on top of the rubble (clay and shell debris) brought in by lorries to fill in the pools, as in the case of the new extension to the University of Nouakchott, due to open in December 2023.

5.1.3.3 Landscape features of the Tevragh Zeina ponds

Maximum water depth measured on average over 3 pools: O to 1 m with brackish water.

Soil at depth: sandy-clay, with saline sandy mud at the edge of some pools, grass beds in 2 pools.

Soils on the periphery: apparent salt concentration only in summer and less in the "cold season".

Highest average conductivity, measured in November 2023 in 3 pools: 350 µs/cm, with temperatures ranging from 27 to 30 degrees.

PH: The average pH varies between 8 and 9,

Slightly sloping banks: 0 to 0.3% on average,

Anthropogenic waste: abundance of rubble in the ponds inventoried.

Aquatic life: dragonflies and butterflies permanently above the surface of the water during the day, spiders, abundance of grass beds,

Biodiversity: moorhens at F Nord, redbud, white heron, turtle doves.

Invasive species: Typha in ponds on F Nord and in small copses in certain fences, sessuvium,

Observed use of the ponds: bathing for street children, refreshing of stray dogs,

Sanitation: ONAS pumping station at F Nord.

6 SUMMARY OF AVERAGE MEASUREMENT DATA (PH, T, C) FOR THREE DIFFERENT PONDS IN THE COMMUNES DE FRONT DE MER (EL MINA, SEBKHA, TV ZEINA)

We chose one pond per commune on the basis of similarities and took measurements of PH, T and conductivity for each site. We then took monthly averages for the three ponds. In terms of PH, **it** rose to 8, peaked at 9 in December and then fell back to 8 again. This is probably due to the subterranean influence of high water along the coast, which has been rising steadily, preventing fishermen from taking to the sea on 13, 14 and 15 January 2024.

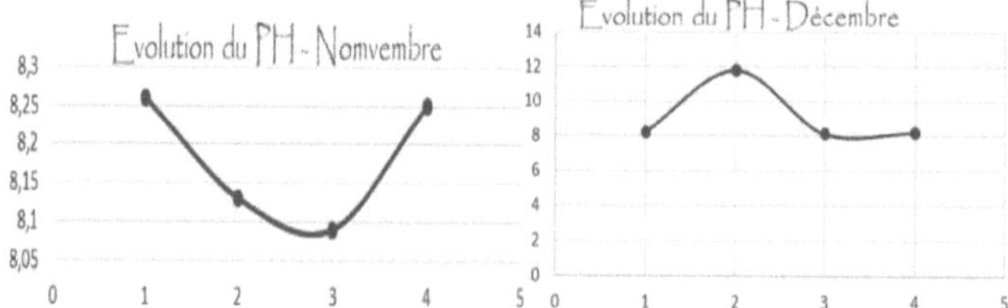

Temperatures fell sporadically over 3 successive days in November 2023, then over 5 discontinuous days in December 2023, before dropping again over 7 days in January 2024.

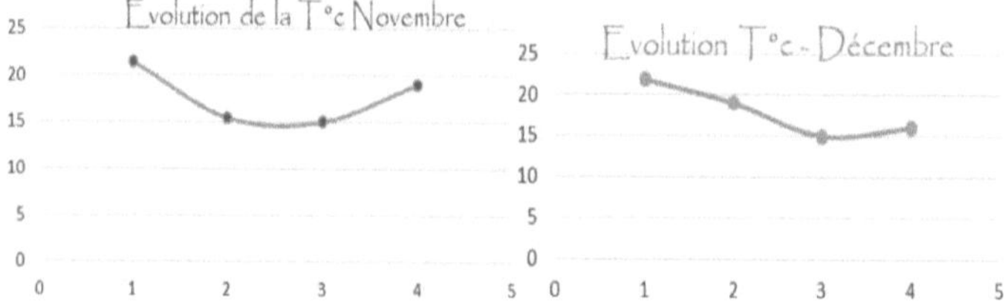

Conductivity was certainly influenced by average temperatures, but our assessments are affected by the high volatility of the conductivity value. It fluctuated between 105 and 125 µs/cm (Hanna device).

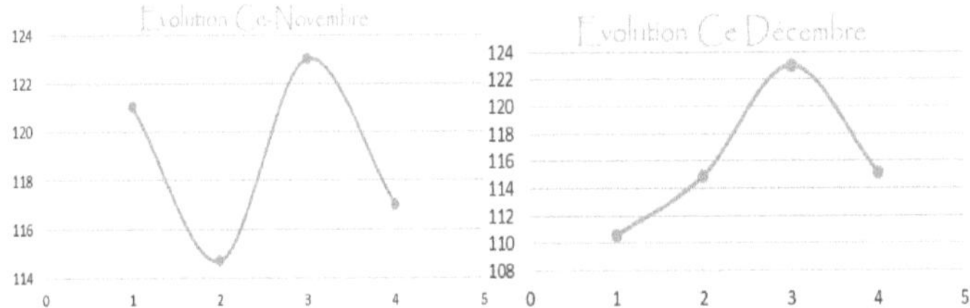

In any case, these data remain indicative, and must be confirmed by the probe still to be ordered by the WACA project.

7 TYPOLOGY OF PONDS IN NOUAKCHOTT AT THE FRONT OF CONTINENTAL DUNES

These three communes, so named because of their geographical position, are built on or at the edge of continental dunes. These sandy areas are part of the ancient morphology of the Nouakchott area (Ogolian dunes and Erg Amoukrouz) but have been reshaped by urban pressure and erosion. These dunes run in an East-North-East direction and are separated by depressions whose altitude varies from one commune to another. For example, the very low depressions in Dar Naim vary in altitude from -3m to 4m, which makes this commune more vulnerable to the proliferation of pools caused by rainfall, or to the rising salty water table. This is not the case in Teyarett, where average altitudes range from 1.5m to 6m. This does not prevent the inter-dune areas of this municipality from having altitudes of 0 to 3m. As a result, ponds abound in Dar Naim, making it the municipality with the highest number of inventoried ponds in Nouakchott. The continental dunes have tilted in favour of the communes to the east of Nouakchott, such as Arafat and Toujounine, where ponds are virtually non-existent, with altitudes ranging from 2 to 11 metres in the latter commune. However, for reasons of importance and geographical distribution of ponds in Nouakchott, we have classified the Toujounine commune with the group of communes known as "Communes with localised pools: Ksar, Riadh and Toujounine" (Chapter 8).

Name of the municipality	characterized as	Surface area in Km2	Population 2021
ARAFAT	Urban	38	186 000
DAR NAIM	Urban	20	147 000
TEYARET	Urban	18	81 000

7.1 TYPES OF PONDS IN THE COMMUNES BORDERING THE CONTINENTAL DUNES: ARAFAT, DAR NAIM AND TEYARETT.

These communes are located at the western and central (Arafat) limits of the continental dunes of Nouakchott. The outcropping of ponds (particularly widespread in Dar Naim) is the result of physical vulnerability, aggravated by very dense urban occupation and intensive sand extraction. In practical terms, this means that materials are extracted for building purposes in some places, and filled in others. The dramatic consequences in these communes are the lowering of the water table, added to the fact that the extraction areas are transformed into ponds and areas of stagnant water as a result of heavy rain and rising capillaries. In the series of photos below, you can clearly see the process of transforming dune areas into ponds in the so-called continental communes of Nouakchott:

Photos 15 in series : PROCESS OF "MARISATION" OF DUNE AREAS IN THE MAINLAND COMMUNES OF NOUAKCHOTT :

1: Dune coveted for its sand2: urban pressure3: perennial outcropping pond

Tarhil before material extraction 2004/ Appearance of waterhole and general outcrop 2010/ Urban pressure waterhole 2021

7.1.1 Typology of ponds in ARAFAT

Ponds with brackish water are few and far between in the commune of Arafat, although temporary pools do appear during the heavy rainy season, and are concentrated in low-lying areas with a water level of 0.5 m, such as in the Market area and post 11.

7.1.2 Preliminary inventory of ponds in Arafat November 2023

The teams identified two perennial ponds in Arafat in November 2023, but people still say that everything is a pond, especially during the rainy season.

Map 7: Preliminary inventory of ponds in Arafat in November 2023 (source: Leerg on Waca 2023/2024 research programme)

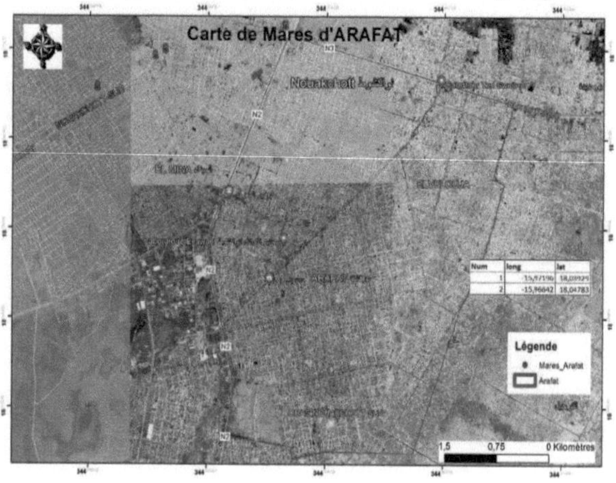

7.1.3 Profile of ponds in Arafat

As mentioned above, this is what people in Arafat distinguish as ponds in the rainy season:

Photo 16: Flooding in Arafat following the 2022 rainy season (source: Author 2023)

Flooding in Arafat in August 2022.	Wintering in Arafat in 2022

7.1.3 Landscape features of the Arafat ponds

Maximum water depth measured on average in the 2 ponds inventoried:
O to 1 m with brackish water.

Depth of seasonal pools at Arafat: 0.2 m

Soil around ponds: sand and clay with shells,

Highest average conductivity, recorded in November 2023 in the 2 pools:
280 µs/cm, with temperatures varying between 28 and 31 degrees.

PH: The average pH varies between 8.5 and 9,

Anthropogenic waste: abundance of rubble and waste in the shallows of ponds.

Aquatic life in the 2 ponds: None

Biodiversity: frogs reported by several people,

Invasive species: Waterloti around ponds,

Observed use of the ponds: bathing for street children, refreshing of stray dogs,

Sanitation: tanker operations during winter storage.

7.1.4 Typology of ponds at DAR NAIM: LOWLAND AND BETWEEN 2 DUNARY SLOPES

Dar Naim's exceptionally high number of ponds in Nouakchott is due to several factors:

- The first factor lies in the morphology of Nouakchott itself, where the Dar Naim area is located in a geographical area defined in local toponymy as a "Teyarette", which means a flat "Gawd" between two sandy mounds. This Teyarette is the northern extension of the site of the old airport, and therefore the wells or oglats of Nouakchott in the past.

- The entire low-lying area of Dar Naim, from the old airport in the west to the north of the "Ould Badou" crossroads, is at an altitude of between 0 and 2 metres above sea level. This was the area of Nouakchott's ancient watering holes, as described by the French in 1903, when they created the Nouakchott post: "From the sea... we climb a high dune stabilised by reddish sand and descend to **a valley with a whitish bottom** (a Teyarette) where the wells of Nouakchott are dug. These are rudimentary excavations where water stagnates, in small quantities, at a depth of three or four metres" (According to people interviewed who lived through the creation of the Nouakchott post, the last well (oglat) was 3 metres deep, on the edge of the northern perimeter of the old airport fence, on the outskirts of Dar Naim). In 1923, Théodore Monod also said: "Several of the oglats (wells) have collapsed and are used as drinking troughs by warthogs and jackals... as soon as you dig a well, the water appears in a level where shells abound".

- The two sandy mounds to the east and north-west encourage subterranean infiltration into the central basin of Dar Naim. What's more, the vegetation that stabilised these dunes and regulated infiltration through plant roots has completely disappeared under the impact of urbanisation. Even the green belt built in the 1980s has not been spared, and has now virtually disappeared.

- Finally, from 1983 to 2003, all the low-lying areas of Dar Naim were subject to uncontrolled extraction of shellfish and sand for construction purposes, which led to the continuous raising of the water table. And with the slightest rain, this water table encourages the formation of pools and puddles, which are difficult to clean up.

7.2.2 Preliminary inventory of ponds in Dar Naim November 2023

Unlike the other communes, the inventory of ponds in Dar Naim took three weeks to complete, given the general outcrop of ponds throughout the commune. Some people thought that there were many more pools in the seafront communes (El Mina, Sebkha and Tevragh Zeina), but this was without taking into account the very low topography of Dar Naim, and the disfigurement of its landscape by sand and shell quarry vendors, which led to more than a hundred pools being inventoried. On the inventory map of the ponds in Dar Naim, you can see the large 10-hectare pond with its year-round perennial water, and the string of other secondary ponds to the north-east of the former airport. People have come to terms with this marshy landscape, although some households have simply abandoned their homes or concessions, which have been invaded by the water. The Dar Naim municipality estimates that 350 families abandoned their homes in 2022 alone.

Photos 18 in series: Traces of the fences of houses abandoned to salt and rubbish in Dar Naim (source: Author 2023)

Location: Ponds 18 to 25 in Dar Naim

➤ Town name : Dar Naim

➤ Geographical coordinates: Latitude: 18:5:49.92

Longitude: -15 :56 :15,936

➤ Date : 17/11/2023Time : 10h15

➤ Leerg team name: Dar Naim team

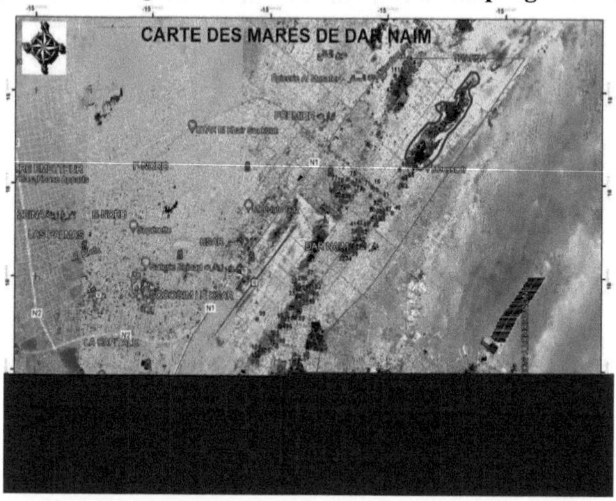

Coordinates of the ponds at Dar Naim, which contain the largest number of ponds in Nouakchott, even though it is not a coastal municipality:

Num	Long	Lat
1	-15,92287	18,1349
2	-15,91708	18,13571
3	-15,91777	18,13649
4	-15,91368	18,13955
5	-15,91523	18,13567
6	-15,91538	18,13591
7	-15,91555	18,13647
8	-15,91573	18,13507
9	-15,91712	18,13294
10	-15,91708	18,1328
11	-15,91703	18,13813
12	-15,91525222	18,14277722
13	-15,91477	18,14074111
14	-15,91505	18,14034
15	-15,91703	18,13813
16	-15,92197	18,11655
17	-15,93769	18,09376444
18	-15,9377	18,09371
19	-15,9383	18,09316
20	-15,93807	18,09402278
21	-15,93905	18,09384
22	-15,93843	18,09411
23	-15,93837	18,09486
24	-15,93817	18,0956

25	-15,93776	18,0972
26	-15,93855889	18,09777
27	-15,93609	18,10149
28	-15,92833278	18,10442333
29	-15,92822	18,1034
30	-15,92786	18,10365
31	-15,92802	18,10319
32	-15,92765	18,10397
33	-15,92802	18,10319
34	-15,92676	18,10398
35	-15,92672	18,10431
36	-15,92538444	18,10492
37	-15,9259	18,10474
38	-15,92679	18,10539
39	-15,92475444	18,10781
40	-15,92518	18,10637
41	-15,925	18,10835
42	-15,92474	18,10729
43	-15,92441	18,10861
44	-15,92433	18,10824
45	-15,92256	18,10973
46	-15,922	18,11042
47	-15,91708	18,13571
48	-15,91743333	18,11157
49	-15,92048	18,11203778
50	-15,92066	18,11201
51	-15,92002	18,11407
52	-15,92204889	18,11248
53	-15,93609	18,09475
54	-15,93679	18,09484
55	-15,93712	18,09529
56	-15,93702	18,09561
57	-15,93646	18,09889778
58	-15,9361	18,0955
59	-15,93601	18,09605
60	-15,93531	18,09628
61	-15,93504	18,09613
62	-15,93483	18,0962
63	-15,91593	18,11651
64	-15,91563	18,11712
65	-15,91450778	18,11857389
66	-15,91593	18,11987
67	-15,91371	18,12085

68	-15,91321	18,12166
69	-15,91395	18,12205
70	-15,92227	18,11492
71	-15,92406	18,11623
72	-15,92715667	18,11673
73	-15,92579	18,11707
74	-15,92618	18,11737
75	-15,92728	18,11805
76	-15,92936	18,1172
77	-15,9301	18,1172
78	-15,92991	18,11675
80	-15,93053	18,11585
81	-15,92975556	18,11495
82	-15,928905	18,11391
83	-15,92847	18,1137
84	-15,92761	18,11366
85	-15,92075	18,11007
86	-15,92013	18,11075
87	-15,9194	18,11122
88	-15,93769	18,09218
89	-15,93831028	18,09156
90	-15,93885	18,09148
91	-15,94032194	18,09161
92	-15,94046	18,09212
93	-15,94137333	18,09080444
94	-15,94133	18,09074
95	-15,94118	18,09052
96	-15,94105	18,08993
97	-15,94156	18,08992
98	-15,94252889	18,08567
99	-15,94688	18,09092444
100	-15,94766778	18,08284
101	-15,94237	18,08776
102	-15,94345333	18,08765
103	-15,94358	18,08711
104	-15,94601	18,08445833
105	-15,94648	18,08385
Grande Mare	-15,905151	18,13113

7.1.5 Profile of the ponds at Dar Naim

The south-western area of Dar Naim, which in the years of independence was known as the "Well Zone of Nouakchott", has now been transformed into densely populated neighbourhoods with pools of water for which there is no solution, let alone a connection to the sewage network. As a result, some houses have been abandoned, or households forced to flee by the intrusion of water take refuge in other neighbourhoods, at least for the duration of the rainy season. The precariousness of Dar Naim's neighbourhoods, surrounded by ponds, is striking, and the risks run by the population include electrocution, foul odours, water-borne diseases, etc. Faced with this situation, the local authorities have no resources, especially in terms of sanitation, despite the State's support in the form of ONAS tankers during the rainy season. During the inventory, the people living near the ponds in Dar Naim gave us their own profile of the ponds, which nevertheless have their own realities on the ground. According to the households interviewed, there are :

- House ponds, which are located in the courtyards of houses, and whose water level is at its highest during the rainy season.
- The puddles in the alleyways are mitigated by the residents of the neighbourhoods, by placing large stones on the surface of the water, in order to facilitate pedestrian traffic between neighbourhoods. However, these households deplore the inaccessibility of their homes during the winter months by ordinary means of transport, even carts.
- The large ponds defined (particularly on the edge of the old airport) by the local population as ponds that never dry up, no matter how much the State and the municipality fill them in. These ponds are a haven for everything: stray dogs, rubbish dumps, salts, escape routes for delinquents, and even shelters for street children, etc.

According to ONAS, the pumping and backfilling have had a positive effect, drying up some neighbourhood ponds and reducing the level of septic tanks, as in other Nouakchott communes, particularly those on the seafront.

7.1.6 Landscape features of the Dar Naim ponds

Maximum water depth measured on average over 5 ponds inventoried: O to 0.7 m with brackish and slightly brackish water.

Depth of wintering ponds at Dar Naim (November 2023): 0.1 m

Soil around pools: shell clay to sandy clay; mud in some pools

Highest average conductivity, recorded in November 2023 in the 5 ponds: 250 µs/cm, with temperatures varying between 28 and 32 degrees.

PH: The average pH varies between 8 and 9,

Anthropogenic waste: abundance of rubble and household waste in the shallows of ponds,

Aquatic life in the 5 pools: Algae observed in some pools,

Biodiversity: none

Invasive species: Typha dead in 2 ponds

Observed use of ponds: stray dogs,

Sanitation: tanker operations during winter storage.

It should be noted that the bottom of most of the ponds in Dar Naim is clay with muddy edges. It was also noted that the further you go towards the Dar Naim dunes, the less salt blisters there are. Finally, lithic material was found in the area of the large pond.

7.1.7 Typology of ponds in TEYARETT

Teyarett is a continuous extension of the ancient Ksar area, which with urban growth has seen an influx of a heterogeneous population concentrated first in the Gowd and then spreading towards the northern dunes. The low-lying areas of the commune adjoin the old airport wall fence to the north, almost limiting themselves to the boundaries of the so-called industrial estates. A variety of ponds can be found within these private industrial fences, with sand and clay intercalations. In the north-western part of the municipality, there are continental dunes that have been heavily compacted by vehicle traffic, leaving little room for the emergence of ponds.

7.1.8 Preliminary inventory of ponds at Teyarett November 2023

The inventory of ponds at Teyarett distinguishes between :

- The ponds in the lower part of the commune, which are filled during the rainy season,
- Inter-dune pools.
- Artificial ponds have also sprung up as a result of the proliferation of sand and alluvial quarries.

The first category of pools is characterised by evaporite deposits with the presence of shell debris and clay. The saline concentration of the water is limited, in comparison with ponds in coastal communities.As for the inter-dune pools, the water is not outcrops, but its underground presence is evidenced by the humidity of the sandy surface layer. As a result, the field teams sometimes found the pools to be misleading.Finally, there are the artificial ponds, which are confined to the areas where unauthorised quarries have been dug in the commune. Their hollows can fill up during heavy rainfall, but with the permeability of the sand, the water quickly infiltrates leaving only a few traces of damp soil.

Map 9: Preliminary inventory of ponds at Teyarett in November 2023 (source: Leerg on Waca 2023/2024 research programme)

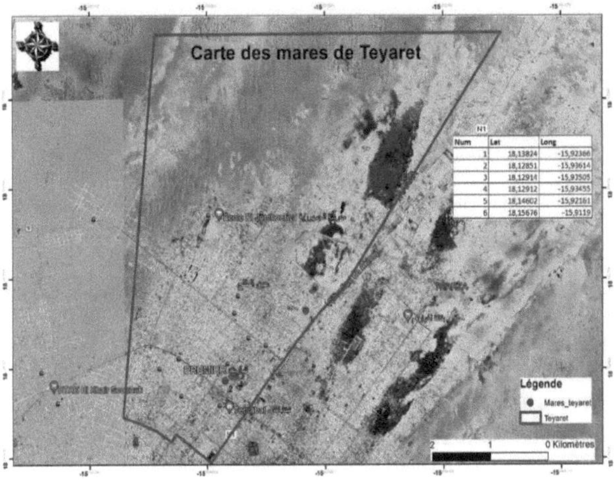
Carte des mares de Teyaret
N1
Num Lat Long
1 18,13824 -15,92366
2 18,12851 -15,93614
3 18,12914 -15,93505
4 18,12912 -15,93455
5 18,14602 -15,92161
6 18,15676 -15,9119
Légende
Mares_teyaret
Teyaret
2 1 0 Kilomètres

7.1.9 Profile of the ponds at Teyarett

The pools at Teyarett are between 1 and 2 metres high, or even 2.5 metres high in the dune peaks. They are generally located in depressions or sandwiched between two dunes. Of course, the morphology of the commune can be used to identify sediments of marine origin in depressions, which are probably the result of marine transgressions. Generally speaking, the depressions in the commune of Teyarett are gradually drying out as a result of urbanisation, wind dynamics and evaporation. The concentration of rubbish in the pools is such that the surface of the water is no longer distinguishable, as in the photo below:

Photo19 : Pond outcropping à Teyarett but submerged by rubbish (source: Auteur 2023)

/Geographical coordinates: Laltutide:18:7:39,468 Longitude:-15:56:0,276/Date:05/02/2024. 17h:28

7.1.10 Landscape features of the Teyarett ponds

Maximum water depth measured on average in 2 ponds inventoried: 0.5 m to 0.8 m with slightly brackish water.

Maximum depth of wintering pools at Teyarette (November 2023): 0.1 m

Soil around ponds: sand, clay, alluvial deposits

Highest average conductivity, recorded in November 2023 in the 2 pools: 262 µs/cm, with temperatures varying between 29 and 33 degrees.

PH: The average pH varies between 7.9 and 9,

Anthropogenic waste: abundance of household waste,

Aquatic life in the 5 ponds: No life observed,

Biodiversity: none

Invasive species: sometimes Waterlotti

Observed use of ponds: rubbish dumps,

Sanitation: no action reported by local residents.

8 TYPOLOGY OF PONDS IN NOUAKCHOTT IN COMMUNES WITH CONTINENTAL AND COASTAL MARGINS (TOUJOUNINE, RIADH AND KSAR): SAND AND SHELLFISH

Communes on the continental and coastal margins of Nouakchott :

Name of the municipality	characterized as	Surface area in Km2	Population 2021
RIADH	Urban	23	121 000
TOUJOUNINE	Urban	27	146 000
KSAR	Urban	18	57 000

In Nouakchott, the limits of the continental dunes and their interference with the shellfish marine area are really evident in these three communes. Indeed, these communes have vast landscapes characterised by the presence of shells on dunes and under dunes, and by white, red and beige sands. And between their dune spaces, low-lying areas with variable pools, depending on the topographical levels. The pools in these shallows are particularly characterised by their white or beige dune shores on a sandy-clay base, as shown in the photo below. Because of the specific nature of these pools (which emerge from the dunes), geomorphologists refer to them as **continental and coastal margin pools.**

Photo 20: Tarhil pond with continental and coastal margins

However, in the commune of Ksar, the morphology is very marked by continental clay strewn with shell debris, which used to be the delight of quarrymen when building the urban roads in Nouakchott, a city that was just emerging in 1958.

8.1 TYPES OF PONDS IN MUNICIPALITIES WITH CONTINENTAL AND COASTAL MARGINS: SAND AND SHELLS

At the beginning of certain ponds in these communes, we have the triptych first of all of quarries, which dug for : Sand + shells + everything = **a** definite outcropping pond (S + C + T = M); either by the rising water table or by the accumulation of rainwater.

Photos 21 in series : At the beginning of a man-made pond at In Nouakchott, we exploit : S (sand) + C (shellfish) + T (waste) = FLOATING TIDE when the sea rises seasonally or when it rains for the first time.

8.1.1 Typology of ponds in RIADH

The municipality of Riadh is home to shallow sebkhas and low-lying areas, with alternating brackish water surfaces that vary according to rainfall and hydrostatic factors. The seasonal rise in the water table, the salinity of the water (brackish or salty) and the silting up give the commune of Riadh landscapes typical of an area with continental and coastal margins. These factors result in a topography made up of clay-limestone soils, ponds, sebkhas and dunes, according to the zoning of the municipality. Geomorphologically, the commune is divided into two zones:

- The western zone bordering L'Aftout Essaheli, with sebkhas and low-lying clay and shell areas at an altitude of 0 to 2 metres (photo 20).

- The eastern part, at the edge of the Trarza erg, with dune complexes and clayey lowlands at altitudes of 1 to 3 metres.

Photos 22 in series: with (A and B): Type of waterhole in the boundary zone of L'Aftout(A) and Type of waterhole in the boundary zone of the continental Erg of Trarza(B)

A: B :

At the time, in 1990, to meet the construction needs of the households in the fledgling commune, the very poor inhabitants had no hesitation in clearing as much land as possible, here and there, sometimes even in the middle of the street, in search of the shells and sand they needed to build their homes. Constant digging everywhere resulted in a disfigured landscape in Riadh, with excavated fields and outcrops of inter-dune pools.

Photo 23: Pond with sandy bottom on dune at Riadh :

8.1.2 Preliminary inventory of ponds in Riadh November 2023

The map of the Riadh commune clearly shows the interference between the dunes and the shellfish shallows. The central Tarhil pond is at the heart of the newly urbanised area of the commune. We applied a questionnaire A quick questionnaire was sent out to people living near the waterhole, and a summary of this questionnaire was compiled in order to assess people's perceptions of the waterhole. The households interviewed and living around the waterhole had differing views, with the following opinions:

- "The waterhole must be filled in by the State, and its lots allocated to their beneficiaries".

- "Those who dug the pond must be punished" (allusion to sand sellers)

- "The pond is ALLAH's work, and we all belong to God".

- "The Riadh pond is a rubbish area and it poisons our lives,
especially as the children bathe there every day".

Some of the households questioned had no response.

On the other hand, by questioning a number of informal workers (carters, shellfish sellers, handlers, administrative planters) and their families around the waterhole, we were able to establish the following impressions, cross-checking the observations of users and the public frequenting the waterhole:

THE FEELINGS OF USERS OF THE POND :

With reference to the vision of the mayor of Riadh for the waterhole, and by questioning the users who come to enjoy the coolness of the waterhole (because its imposing body of water attracts a public in the evening, especially on a hot day), it was felt that the Riadh waterhole and its right of way could be developed further, especially for a well-informed public wanting to find an environmental island in the middle of the urban jungle. Some of the users we met at the site wanted a minimum of infrastructure, such as public benches in front of the pond, green spaces and even a fast-food service.

Categorisation of users of the Riadh waterhole: Observed practices of users of the Riadh waterhole:

	In a group
Walking/hiking/hiking	By car and then on foot
	Jogging and exercise as a couple
Sport	Local children play football every afternoon in front of the pond
	Children with bikes
	family football
Sitting in contemplation of the pond	child supervision
	contemplation of the water landscape and the dunes all around
	Some threw stones into the water
	Picnic on the banks of the pond
Around the pond	Tea on the dune beside the pond
	Birdwatching
	Children swimming in the pond
Spontaneous visit	contemplative
	wonder
	Water fans
	with your family
Other	photos with landscape background of the pond

The informal recreational activities around the Riadh waterhole can be categorised as follows. There are essentially itinerant uses, such as walking along the shores of the waterhole, jogging and tea-tasting, particularly at the top of the dunes and on the edge of the waterhole. There are also sports activities and bike rides, particularly for young people from the neighbourhood. So the Riadh pond has a profile linked to a number of leisure and sports activities for the town's residents. It's fair to say that the residents live and breathe their waterhole, unlike the rest of the population of Nouakchott's neighbourhoods.

Map 10: Preliminary inventory of ponds in Riadh in November 2023 (source: Leerg on Waca 2023/2024 research programme):

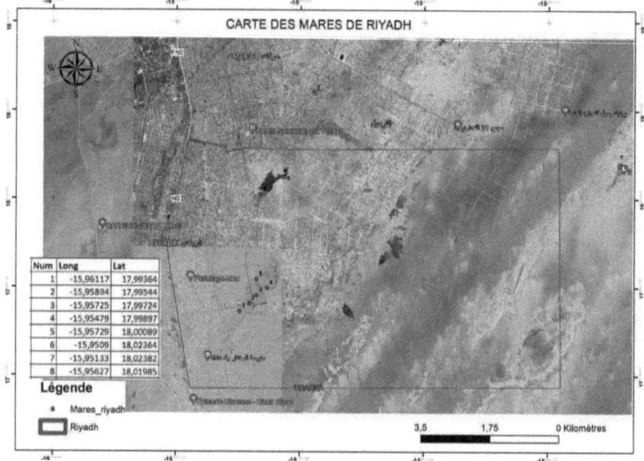

In the middle of the map, you can see the Tarhil pond and the amount of water it contains, in a dune environment.

8.1.3 Profile of ponds in Riadh

The commune of Riadh presents a miniature image of all the types of ponds in Nouakchott: sebkhas ponds, inter-dune ponds, ponds of rainfall origin, etc. This gives the commune several typical profiles of ponds. This gives the municipality several typical pond profiles: **Profiles: A, B, C, D, E, and F and photos 24 in series:**

Profile A: Sectioning of the Tarhil pond to enhance its biodiversity: as for example at F North (in 1 below); or developing Krill in the Tarhil pond (small worms in red in 2 below):

1: Biodiversity pond (the latest news is that the pond has been filled in by property developers/June 2024, unless the government can find them land elsewhere/Farewell to the moorhens).**2: Krill pond**

The Krill (photo in 2) developed in the experimental pond-leg, can favour the residence of seasonal birds, as well as other aquatic species. Krill refers to a group of small living organisms (shown here in red in the photo at the bottom of the pond) that provide food for many aquatic species, including sea fish and birds. Some marine biologists consider krill to be one of the bases of the food chain in brackish waters. It could therefore be a fundamental component of ecosystems for the biodiversity of certain ponds in Nouakchott.

Profile B: Pond abandoned by shopkeepers and inconvenience for city dwellers in Riadh :

Profile C: Pond polluted but used for bathing by children in Riadh and with their happiness:

Profile D: Pond pumped since 2020 without end, and headache for ONAS and the military engineering, which have pumped so much to the sea and it continues (in the background the motor pumps installed by the two institutions):

Profile E: Garbage ponds and a housing estate in Tarhil: Waiting for the water level to drop and for contractors to start building (in a high-risk area):

Profile F: Pond with halophilic vegetation and bird resting places seasonal workers or those migrating to Tarhil :

8.1.4 Landscape features of the Riadh ponds

Maximum water depth measured on average at 2 ponds inventoried :
O to 0.8 m with brackish and slightly brackish water.

Depth of wintering ponds at Riadh (November 2023): 0.5 m

Soil around ponds: sandy to clay-shell soils

Highest average conductivity, recorded in November 2023 in the 5 ponds: 230 µs/cm, with temperatures varying between 29 and 32 degrees.

PH: The average pH varies between 7.8 and 8,

Anthropogenic waste: household waste and used oil in the shallows of ponds,

Aquatic life in the 5 ponds: seabirds and Krill in pond 9

Biodiversity: halophilous plants, dragonflies, butterflies, lizards,

Invasive species: Waterlotti fairly present but not widespread

Observed use of ponds: stray dogs, cats, children bathing, escapes around ponds in hot weather.

Sanitation: ONAS pumping since 2019, with continuous stoppage in November 2023. The ONAS agent we met on site did not give any figures on the volumes pumped or the results. However, in 2015, the Ministry of Hydraulics provides the following data for pumping at the Atoit pond between Sebkha and Tevragh Zeina:

- Volume of water pumped from the ATOIT pond in 2015: 6 million m3.
- Evacuation, 3 electric pumps (flow rate 100m3/h and HMT 60m), initially planned for pumping at the Ksar, have been installed.
- Replacement of the electric pumps with four more powerful 240m3/H - HMT 60m pumps.
- Pumping 5 hours a day except at weekends,
- As a result, the level of the pond fell by 0.5 m, but when pumping was stopped for 11 days due to technical problems, the level of the pond rose by 0.1 m.

In terms of water surface area, the Atoit pond is similar to the Tarhil pond. It should be noted that the bottom of three ponds in Riadh is made up of sand.

8.1.5 Typology of ponds in TOUJOUNINE

Ponds are almost non-existent in Toujounine due to its geographical location and dune relief. However, there are depressions (Gowd) such as in the valley running from the Route de l'Espoir to the ground station. These depressions are surrounded by dune massifs aligned parallel to the prevailing winds (north-east, south-west). With the intense wind activity at Toujounine, the depressions expose the bedrock, which becomes waterlogged during the winter months, but does not generate any significant pools. At low points in the subtratum, shell deposits left by marine transgressions in the Quaternary can be seen, as well as sandy-clay surface layers.

8.1.6 Preliminary inventory of ponds in Toujounine November 2023

This Google 2014 photo clearly shows one of the low-pressure areas at Toujounine in the area of the earth station. During heavy rains, this depression can generate seasonal pools.

Map 11: Preliminary inventory of ponds in Toujounine in November 2023 (source: Leerg on Waca 2023/2024 research programme)

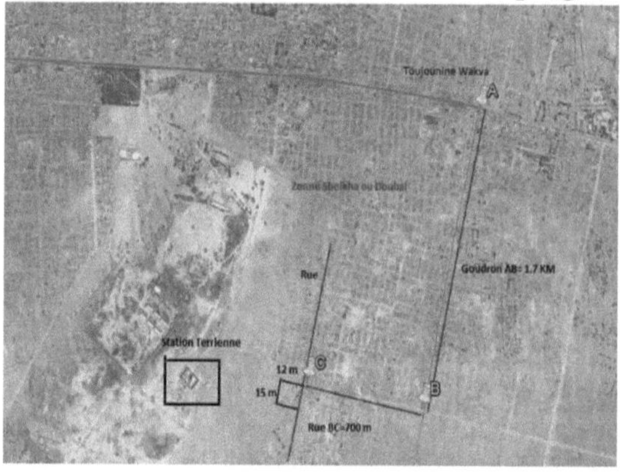

It should be remembered that Toujounine is the most continental of Nouakchott's communes, and when the sea flooded Chott Boll in December 1984 (rising as far as 63 km from Nouakchott through the Essahéli Aftout), the governor of Nouakchott at the time planned to relocate the populations of El Mina and Sebkha. The commune also has the highest altitudes, with levels of up to 5 m (see map of altitudes in Nouakchott at the beginning of this report. Today, with Tarhil, the municipality is set to welcome several families, which will further increase its urban population.

8.1.7 Profile of the ponds at Toujounine

The profile at Toujounine is more likely to be a profile of the inter-dune depressions, which requires further field investigations in this commune.

8.1.8 Landscape features of the Toujounine ponds

In the landscape of the Toujounine commune, the inter-dune depressions need to be monitored, especially in the event of exceptional flooding and in the context of climate change. If monitoring is to be carried out, it will be necessary to determine whether the Toujounine depressions have undergone lagoon phases. or lacustrine soils. The presence of Zygophyllum waterlotii in the Gowd of Toujounine is a sign of the presence of halomorphic soils, which are generally under the effect of the salt wedge. This plant also proliferates at dune level, forming small mounds in its vicinity.

8.1.9 Typology of ponds at KSAR

The Ksar is the founding embryo of Nouakchott and was erected in 1903, with the creation of the town. However, due to the location of the town's core, in the middle of an inter-dune basin, the urban agglomeration was destroyed by the river floods of 1950/1951. The traditional wells of the city of Nouakchott are dug in this basin near the current "Mother and Child Hospital". The SNDE, which currently manages the concession containing these wells, is having difficulty controlling the groundwater emanating from these excavations. The main ponds in the commune are located around this area, and in the grounds of the old airport. Due to urban density, many ponds are submerged by installations linked to economic and industrial activities (garages, factories, etc.). In the area known as the garages, ponds alternate with mechanical waste and oil spills. The teams found it difficult to clearly distinguish the outlines of the pools.

8.1.10 Preliminary inventory of ponds in the Ksar November 2023

On the map, the upper Ksar is made up of dune areas with villas, housing and administration, notably the Château d'Eau districts and the sites of a number of projects. To the west of the old airport lies the lower Ksar, with altitudes ranging from 0 to 1.5 metres.

Map 12: Preliminary inventory of ponds in the Ksar in November 2023 (source: Leerg on the Waca 2023/2024 research programme):

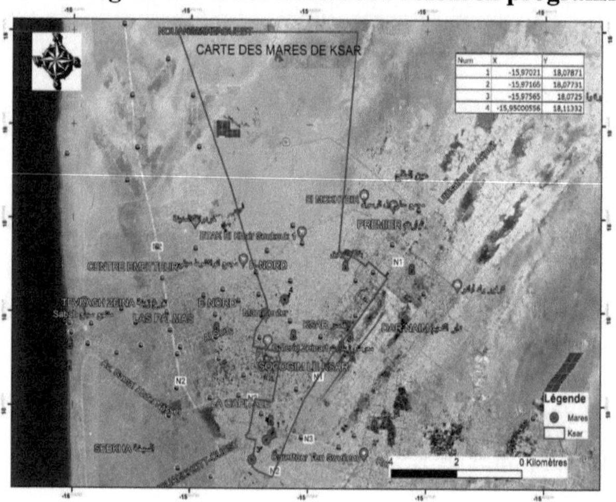

8.1.11 Profile of the ponds in the Ksar

Several pools in the Ksar have been filled in or evacuated, either by the military engineers, the ONAS or the Chinese project. However, despite all these efforts, some ponds have reappeared after all these actions, notably the ponds bordering the Ksar and Dar Naim, as well as in other areas, notably in Tevragh Zeina, in the large pond known as the embassy pond (Photo below).

Photo 25: Pond reappearing in t h e embassy zone in Nouakchott after Chinese pumping and ONAS April 2024 :

8.1.12 Landscape features of the ponds in the Ksar

Maximum water depth measured on average in 2 ponds inventoried: O, 5 to 0.8 m with brackish water.

Depth of wintering ponds at the Ksar (November 2023): 0.8 m

Soil around ponds: clay soil with sand content

Highest average conductivity, recorded in November 2023 in the 2 pools: 225 µs/cm, with temperatures varying between 28 and 32 degrees.

PH: The average pH varies between 7 and 9,

Anthropogenic waste: garage rubbish and used oil in the shallows of ponds,

Aquatic life in the 2 ponds: none

Biodiversity: Tamarix, calotropis,

Invasive species: Typha in Baghdad ponds(Socogim)

Observed use of ponds: rubbish dumps

9 LAND ISSUES AND ADAPTING BUILDINGS ON STILTS IN NOUAKCHOTT

9.1 Land ownership of ponds in Nouakchott

A resident of Nouakchott used to say that "ponds defend themselves", alluding to the fact that in areas where ponds outcrop and whatever the State's land allocations, very few people dare to build there. And those who have dared have bitten their fingers off, with salt rising in their homes. The decree implementing the 1983 land law does not specify ponds and sebkhas. However, there is an example to be cited in the commune of Riadh for the Tarhil pond: seeing that the pond was emerging and never-ending, and given that the local people were unable to access their allotted plots before the pond began to outcrop, the Ministry of Housing allocated this pond for communal use in Riadh, and allotted the local people other plots in other areas.

9.2 Buildings adapted to ponds

The example of the construction of the US embassy in Nouakchott, in the large pond on the K Extension block, is worth noting: in the photo on the left, you can see the outlet for collecting underground water which, once it has built up hydrostatic pressure under the foundations of the embassy, is immediately evacuated by a system of slopes towards an open-air pond in front of the embassy premises.

Photos 26 in series : Us embassy in Nouakchott and Adaptation of its construction on ponds and sebkhas, with outlet, each time, the hydrostatic level of the sea, rises.

And photo: Traditional construction on a pond in Benin, and Nouakchott one day? If sea levels rise in the context of climate change.

CONCLUSIONS AND RECOMMENDATIONS

This document highlights the lack of data and studies on ponds, sebkhas and low-lying rainwater flooding areas in Nouakchott. Hence the need to set up a body as soon as possible? A unit? A section? A unit? A department? An observatory? On these three geomorphological units that threaten the capital, and that are vital to monitor for the Mauritanian coastline, in a context of climate change. The following recommendations must be implemented as part of the Waca project, or any other project known as: Coastal ponds and sebkhas in Mauritania: (this cannot wait for a hypothetical coastal observatory that will take at least 4 years to become operational):

- Technically and scientifically identify **pools of rainwater origin**
and **ponds of maritime origin** in Nouakchott ;
- **A technical study to differentiate between ponds and sebkhas** in
Nouakchott and the surrounding area, in particular **fossil sebkhas** (1908-1951)
and **moving sebkhas,**
- Determine underground exchanges between ponds and sea level in Nouakchott.
- Selecting **ponds for biodiversity** in Nouakchott; urban ecotourism needs, with
a programme of halophilic plants (salt absorption in leaves and branches:
Tamarix, Nitraria, etc.) contributing to climate change (carbon), and
development of Krill to attract seasonal birds, and even phytoplankton.
- Selecting **ponds with reduced PH** (a tank of 12 T per week over a surface area
of 1 ha) on the outskirts of Nouakchott (south), to meet the needs of fish
farming, particularly for poor urban women and coastal women (Ndayatt), who
have been underemployed since fish became scarce in the capital.
- Launch a programme to **identify households that have built on ponds and
sebkhas at risk,** particularly and urgently in El Mina, Sebkha, Dar Naim and
Tevragh Zeina-ouest.
- Count the number of **citizens in Nouakchott who have lost their houses** or
fences because of the ponds (climate refugees in their country's capital).
- **Develop the professions of salt producers within the** framework of small
employment enterprises for the benefit of shell and sand sellers in outlying
districts. Selecting sebkhas for hygienic salt production.
- Adopt a **permanent and seasonal comparative inventory programme
ponds in Nouakchott,**
- Identify ponds that have already been filled in and install piezometric

monitoring of

groundwater and its **destination after backfilling,**

- Draw up a **map of high-risk pools** in Nouakchott.

- Weigh up the pros and cons of the developments planned along the coastline, in particular private projects, ports and pontoons, as well as the infrastructure planned by SALN and Meridian.

- Study the long-term influence of :

. The Ndrahamcha sebkha to the north of Nouakchott on the capital's new airport,

. The Saint Louis breach will saturate the soils of the Aftout Essahéli depression and possibly cause flooding towards Nouakchott.

- Determining **the** vital space of ponds and their local impact, in the face of urbanisation.

- Identify the urban areas of Nouakchott with semi-permanent ponds and marshes for **construction on piles** as pilot projects,

RECOMMENDATIONS ADOPTED BY THE MAYORS AND PARTICIPANTS AT THE WORKSHOP ON 15/5/2024 IN NOUAKCHOTT

I. Proposed recommendations:

To develop the study of ponds, with the involvement of other researchers, in order to clarify the origin of water in ponds and their expansion:

1. Studies on coastal sebkhas, particularly those in Nouakchott, and their relationship with seasonal and non-seasonal pools.
2. Map more ponds, and characterise them, particularly in terms of their water function,
3. Study solutions for developing and enhancing ponds,
4. Relocate and compensate people living in flooded or at-risk areas.
- Geo-referenced census of households living in flood-prone areas.
- Census of households that have already abandoned their homes or fences (for the poorest) and have migrated to other municipalities with a continental morphology (climate refugees).
- Declare areas at risk to be State property and
unsuitable for buildings unless they comply with standards.
5. Ban sand and shellfish extraction within the perimeter of the city of Nouakchott,
6. Adopt a consensual GIS programme for monitoring ponds (with a permanent inventory of ponds)
7. Study the influence of the sebkhas, in particular: Ndrahamcha and its risks for the new Nouakchott airport, and the effects of the Saint Louis breach in the event of submersion through the Essahéli Aftout depression.
8. Dissociate the profile of the sebkhas from the profile of the ponds in Nouakchott,
- Assess the risks to all buildings on the Nouakchott coastline,
- Manage pressure on the coast by integrating the climate dimension.

SOME BIBLIOGRAPHICAL REFERENCES

- Fadel M. (2024). Gouvernance littorale et changements climatiques en Mauritanie, Geography - University of Lille, NNT: 2023ULILA020, HAL id: tel-04540412, https://theses.hal.science/tel-04540412. PhD Thesis.
- Ministry of the Environment and Sustainable Development (2017). Plan Directeur d'Aménagement du Littoral Mauritanien (PDALM), partie d'aménagement et prescriptions et annexes [diagnosis and multi-sector investment plan for the Mauritanian coast (2018-2022)].
- Waca project studies, Ministry of the Environment.
- Vernet R. & Tous P. (2004). Les amas coquilliers de Mauritanie occidentale et leur contexte paléo environnemental (VIIe-IIe millénaires BP, Préhistoires Méditerranéennes, doi: https://doi.org/10.4000/pm.111.
- LEERG studies (laboratory for environmental studies and geographical research) University of Nouakchott 2021/2024.
- Experimentation with halophilic plants in the ponds and coastal strip of Nouakchott, INEM(Environmental environnemental de Mauritanie),2022/2023.
- Spécial Littoral mauritanien, special issue of the Mauritanian journal of environmental studies and geographical research.
- Mission Gruvel-Chudeau 1908 from Saint Louis to Port Etienne.

APPENDIXES

APPENDIX 1

PROFIL DES MARES A NOUAKCHOTT :

PHOTO aux 4 coins de la mare/Mare communale numéro-

1/Localisation :

- Nom de la commune :
- Coordonnées géographiques :
- Date : —Heure—/Nom de l'Equipe LEERG—

2/Cochez l'origine de la mare :

- D'origine Marine :
- D'origine pluviale :
- Sebkha :
- D'origine anthropique par enlèvement de carrière de coquillage ou de sable :
- D'origine fuite réseau assainissement :
- D'origine réseau eau SNDE :
- Mare qui s'est créée suite à des travaux publics (construction de routes, etc.) :
- Autre :

3/Description du site de la mare : entre deux dunes, en bas-fonds, etc.

4/Information locale sur la Mare, ou auprès de la commune :

- Remblayée par le génie militaire :
- Ou remblayée par la commune :
- Ou bien par un promoteur immobilier privé :
- Ou bien Mare en cours de pompage par l'office d'assainissement (présence de motopompe) :
- Ou bien construction de maisons sur le site de la Mare :
- Ou bien Mare traversée par une route goudronnée ou piste en tout venant :
- Autre :

خصائص و مميزات المستنقعات في نواكشوط:

خذ صورة من الجهات الاربعة للمستنقعات

١/الموقع:

- اسم البلدية:................
- الإحداثيات الجغرافية(خط الطول و العرض):...........
- التاريخ،الساعة،إسم الفريق:..............

٢/ضع العلامة على الجواب الصحيح:

- مستنقع بحري
- مستنقع مطري
- مستنقع على اساس سبخة
- مستنقع ناتج عن حفر من اجل التراب او المحار
- مستنقع ناتج عن تسريب في شبكة الصرف الصحي
- مستنقع ناتج عن تسريب مياه SNDE
- مستنقع ناتج عن اشغال عامة او بناء طرق
- أخر

٣/يوصف لموقع المستنقع:بين كثبان او منخفض

٤/معلومات محلية حول المستنقع او كملها لدى مصالح البلدية

- دعم من طرف الهندسة العسكرية
- دعم من طرف البلدية
- مستنقع ناتج عن أشغال خصوصية
- مستنقع يُخطط حاليا من طرف مكتب الصرف الصحي (إذا كان هناك ألية دفط المياه)
- منازل و دور على موقع المستنقع
- مستنقع معبور من طرف طريق الإسفلت أو مسار معبد
- أخر

5/Paysage visuel de la mare :

- Infestée d'ordures :
- Couverte en végétation :
- Mare toute couverte de sels :
- Mare dont la surface est totalement couverte en eau :

6/Exploitation humaine de la mare :

- Extraction du sel :
- Mare fréquentée par les enfants du quartier en baignade :
- Mare souvent utilisée comme aire de récréation et de contemplation paysagère pour citadins et ménages du quartier en mal de Nature :

7/Situation foncière :

- Mare lotie et abandonnée :
- Mare située en place publique :
- Mare en concession industrielle privée :
- Mare désignée pour abriter des bâtiments publics (par exemple, écoles, dispensaires, etc.) :
- Autre :

٥/منظر المستنقع:

- موبوءة بالقمامة
- مغطات بالنباتات
- مغطات بالقشور الملحية
- مستنقع مليئ بالماء

٦/الاستخدامات البشرية للمستنقع

- استخراج الملح
- مستنقع مستخدم للسباحة من طرف أطفال الحي
- مستنقع مستخدم احيانا من طرف الجوار و السكان للتأمل في المناظر الطبيعية

٧/الوضع العقاري للمستنقع

- مستنقع مخطط للعمران و ترك
- مستنقع واقع في مساحة عمومية
- مستنقع واقع في مساحة خصوصية صناعية
- آخر

8/Faune aquatique observée en fonds de la mare :

- Amphibiens (comme grenouille par exemple, etc. :
- Libellules :
- Papillons :
- Autres :

9/Faune en bordure de la mare :

- Lézards :
- Vipères :
- Animaux errants (comme chiens, chats par exemple, etc.) :

10/Présence d'oiseaux :

- Oui : écrivez le nombre d'oiseaux que vous avez vu :

- Non : cochez le mot Non

11/Type de Végétation aquatique :

- Tamarix :
- Nitraria :
- Waterloti :
- Sesuvium :
- Autre :

12/ Végétation bordière de la mare :

- Euphorbes :
- Prosopis :
- Autre :

13/Sols de la mare :

- Sablo argileux :
- Argileux et coquillé :
- Sable dunaire :
- Vaseux :

OBSERVATIONS PERSONNELLES DES EQUIPES DE TERRAIN : ——————————————————

NOMS DES ENQUETEURS +Tél : ————————, ————————, ————————, ————————,

٨/ حيوانات بحرية قد تلاحظها في قعر المستنقع:

- برمائيات كالضفادع مثلا
- اليعسوب (الدبيطان)
- فراشات
- أخر

٩/حيوانات على هامش المستنقع

- سحالي
- أفاعي
- حيوانات سائبة كالكلاب و القطط

١٠/وجود طيور

- إذا وجدت في المستنقع طيور حدد عددها

- إذا لم تجد طيور ضع علامة لا

١١/أنواع النباتات البحرية

- الطرفة
- الگُرزيم
- لميلحة
- المتمدد
- أخرى

١٢/النباتات على هامش المستنقع

- الفردان
- اگرون لمحاند
- أخرى

١٣/التربة المستنقع

- رملية طينية
- طينية مع المحار
- كثبات رملية
- متحضر مختلط بالطين

ملاحظات شخصية لفريق الميدان :............

أسماء طلاب الميدان مع أرقام الهاتف-

APPENDIX 2: OPERATIONS TO FILL IN PONDS IN DAR NAIM AND AN MP'S ALERT ON THE PLIGHT OF PONDS IN EL MINA + THE MINISTER FOR WATER VISITING ONAS'S WORK ON PONDS

A2 : الوكالة الموريتانية للأنباء
Agence mauritanienne d'information

انطلاق عملية ردم المياه الراكدة بدار النعيم في نواكشوط الشمالية

مساءً | 6 نوفمبر 2023 4:22

A3 : نائب يحذر من مخاطر صحية لمستنقع بالدار البيضاء
الأخبار (نواكشوط) حذر النائب البرلماني يحي ولد أبوبكر29 نوفمبر, 2023 –

A4 : نواكشوط: وزير المياه يزور بعض نقاط تجمع المياه في العاصمة نواكشوط14/11/2023
نواكشوط، حيث ستباشر الفرق الفنية التابعة للمكتب الوطني للصرف الصحي حملة زار
وزير المياه إسماعيل عبد الفتاح، اليوم، عددا من نقاط تجمع المياه في العاصمة لوضع حد
لهذه الظاهرة

ANNEXE 3 : EXEMPLE DE BASE DE DONNEES POUR UNE COMMUNE

ANNEXE 4 : couverture médiatique atelier 15 5 2024 :

خميس 09:51 - 2024/05/16

APPENDIX 3: EXAMPLE OF A DATABASE FOR A MUNICIPALITY
APPENDIX 4: MEDIA COVERAGE OF THE 15 5 2024 WORKSHOP :

نظم مختبر الدرسات البيئة والبحوث الجغرافية بتعاون مع برنامج "اوكا "التابع لوزارة البيئة عرض حول التقرير المؤقت عن. خصائص المستنقعات في بلديات انواكشوط ،وحضر العرض عمد. بلديات انواكشوط التسعة

في كلمته بالمناسبة شكر الدكتور والخبير البيئي السيد المختار ولد الحسن شركاء المخبر والقائمين

على مشروع "اوكا ".كما تناول الكلام السيد محمد الأمين ولد باب منسق "مشروع تنسيق المناطق

الشاطئية في غرب افريقيا فرع موريتانيا "الكلام وشكر بدوره العمد وممثلين عن كل من وزارة البيئة

والتنمية المستدامة وجيهة انواكشوط وباحثين من جامعات فرنسية كبوردو وليل والكولونيل بارداس

من إدارة الخرائط الجغرافية لمدينة انواكشوط والقائمين على الورشة و مثمنا لعملهم كما أكد حرص. المشروع على دعم هذا النوع من العروض

عمدة الرياض الأستاذ عبد الله إدريس لقى كلمة بالمناسبة شكر بدوره القائمين على الورشة ،كما. تحدث عن اهم المشاكل البيئية التي تعاني منها بلديات انواكشوط

وبعد ذلك تناوب العمد على الكلام وقد اشادوا بأهمية هذه الورشة شاكرين القائمين عليها وعلى الدور

الذي تقوم بيه في مجال البيئة.

يمتد العرض ليوم واحد حضره بالإضافة إلى العمد خبراء وطلاب مهتمين.

"GET READY TO LEAVE FOR NOUAKCHOTT!

yes

I want morebooks!

Buy your books fast and straightforward online - at one of world's fastest growing online book stores! Environmentally sound due to Print-on-Demand technologies.

Buy your books online at
www.morebooks.shop

Kaufen Sie Ihre Bücher schnell und unkompliziert online – auf einer der am schnellsten wachsenden Buchhandelsplattformen weltweit! Dank Print-On-Demand umwelt- und ressourcenschonend produziert.

Bücher schneller online kaufen
www.morebooks.shop

info@omniscriptum.com
www.omniscriptum.com

Printed by Books on Demand GmbH, Norderstedt / Germany